Quick Study for Your Extra Class Amateur Radio License

By Aleksandra Rohde W3JAG

Dale Street Books
Silver Spring, Maryland
Copyright © 2014 by Aleksandra M. Rohde
All rights reserved.
Printed in the United States of America

"Quick Study for Your Extra Class Amateur Radio License" presents the easiest, most straightforward method to study for the Extra Class Amateur Radio License. It is filled with valuable hints, helpful calculations, exam insights and detailed explanations you won't find anywhere else!

This Quick Study manual also contains the complete 2016-2020 Extra Class Question Pool (thoroughly explained and demystified here) as well as THREE additional streamlined variations on the Question Pool to help reinforce your learning faster and more effectively.

Do not delay any longer. Before you know it, this top ticket in Amateur Radio will be yours!

Aleksandra Rohde, W3JAG

How to Use This "Quick Study" Manual

Congratulations on deciding to go for your Extra Class Amateur Radio License, the top ticket in Amateur Radio! With it come the full range of Amateur Radio privileges and the honor of being at the top of the Amateur Radio world! The only thing that stands in the way of you and your top ticket is the Extra Class exam, so let's talk a little bit more about the exam and what it will take to pass it.

To try for the Extra Class License, you must first have earned your Technician Class and General Class Licenses. General Class licensees may upgrade to Extra Class by passing a 50-question multiple-choice examination. MORSE CODE IS NO LONGER TESTED. Exams are strictly administered by Volunteer Examiners (VEs), who have been accredited by the Federal Communications Commission (FCC)-authorized Volunteer Examiner Coordinator (VEC) system. To pass, only 37 out of the 50 questions need to be answered correctly. The exam tests for knowledge of regulatory provisions, operating practices, advanced electronics theory, radio equipment design and safety.

Studying for the exam is made easier because all the possible exam questions are posted in an official Extra Class Question Pool containing a little over 700 questions. The Question Pool is updated every four years by the National Conference of VECs. **This edition of "Quick Study for Your Extra Class Amateur Radio License" contains the Extra Class Question Pool valid from July 1, 2016 to June 30, 2020.** More than that, this "Quick Study" edition explains the organization of the Question Pool and provides helpful study hints you will not find anywhere else. To help you study even more effectively, three streamlined versions of the Question Pool are also included so you can quickly and easily test yourself on what you know. All this is organized into a simple to use three-step study methodology as follows:

1. **Question Pool Overview.** First of all, it is important to take a step back to look over the Question Pool. What are the topics covered? How many questions in each section? This way you can assess your strengths and

weaknesses. This initial review will help you develop effective study strategies, as we will explain in depth later.

2. **Main Study Effort.** The official Extra Class Question Pool is included here but unlike the version online with its cryptic coding, here the coding system is demystified so you know how to use it. Also included here are valuable study hints and insights into the exam not commonly shared, but which will greatly increase your study effectiveness; such as why it is important to become familiar with the entire multiple choice question and not just the right answers.

3. **Final Review.** As you get closer to exam time you need a good tool to quickly review what you have learned. These three streamlined versions of the Question Pool (the first giving just the questions with the right answers; the second giving just the right answers without the questions and the third giving just the questions) are a terrific way to remind yourself of all you have learned...and what could use a little more brushing up. You might just be tempted to skip the main study effort and proceed right to this step. But please don't! There are good reasons NOT to skip the main study effort, which is explained in more detail later.

When it is time to take the exam, locations and schedules for exam sessions can be found at http://www.arrl.org/find-an-amateur-radio-license-exam-session Also posted on the website is important information about what to bring with you (e.g., required documents, permissible calculators, etc.) on exam day. At the very least, a photo ID is required as proof of identity and some sites require pre-registration for a FCC Registration Number (FRN) to be used in lieu of a Social Security Number. A FRN can be obtained at https://apps.fcc.gov/coresWeb/publicHome.do Depending on the exam site, the exam may be free or a minimal fee may be charged.

If for some reason you do not pass the exam the first time, please do not be discouraged. No one wants you to fail. You can retake the exam as many times as it takes. Some sites, depending on available time, will allow you to retake the exam the same day. The types of exams offered and the possibility of retakes is at the discretion of the individual exam sites. Always check ahead so you know what to expect.

Table of Contents

Question Pool Overview

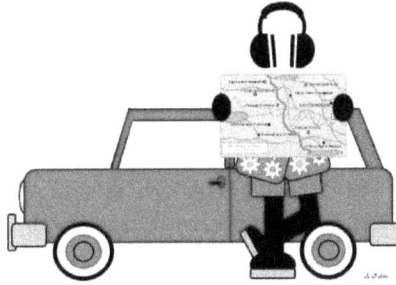

A good way to begin studying for the Extra Class exam is by looking over the Question Pool—kind of like checking your road map before you begin your trip—to get an idea of the distance, travel routes and possible places to stop overnight or for some extra sightseeing. The Question Pool is divided into broad categories called "subelements." Each subelement is further divided into "groups." (The groups are underlined in the Outline provided in this section.) One subelement has only one group while other subelements have six or even eight groups. In total there are 10 subelements and 50 groups in the Question Pool. The important thing to remember is that there will be ONE question on the exam selected from each of the 50 groups, and therefore, 50 questions on the exam.

For example Subelement E1 - Commission's Rules has six groups (A to F). Therefore there will be six questions on the exam from this subelement (i.e., one from every group). So why is this important? Because you want to study smarter, not harder.

Every individual has unique strengths and weaknesses. Some are good at comprehending regulations and others not so much. Some are experts with antennas, or circuitry or RF theory, others not so much. You get the idea. As you review the outline, you start to think about the topics that need more of your attention. This may sound counterintuitive, but unless you have the time to devote, do not spend precious effort trying to master what you do not know—at least not at first.

Play the odds. Remember there are 50 questions on every exam, one question selected from each group. You need to answer 37 out of the 50 questions correctly to pass. Some questions are simpler to grasp while others require more effort. Plus the groups are not all the same size. Some groups have only 11 questions while others may have 17, 18, or even 19! Assume all questions are of similar complexity (they are not—some are easier, others more difficult, but just assume for the time being they are similar). You will spend more time studying the questions in the largest group of 19 than the smallest group of 11. But remember, you will only see one question from each group on the exam, regardless of its size!

So let's say you find one group of questions easier to absorb than another. You will of course want to learn the easier ones first so you have some sure bets under your belt for exam day. Then you can move on to the more challenging ones when you can spend more time. But suppose these more difficult questions are also in the larger groups. This is where time management becomes your friend. Given a limited amount of time, are you going to consume it on a larger pool of questions in a group you are less likely to master? Remember, you only need to answer 37 questions correctly.

The object is to pass, not become Nikola Tesla overnight. The more questions you learn the correct answers to, the greater your chances of success. That is why learning the questions by group or subelement rather than "firehosing" all 700 without a plan is so important. It is a matter of strategy and knowing how to apportion your time and energy. So let's begin with the Question Pool Outline.

Subelement E1 - Commission's Rules [6 Groups]

E1A - OPERATING STANDARDS: FREQUENCY PRIVILEGES; EMISSION STANDARDS; AUTOMATIC MESSAGE FORWARDING; FREQUENCY SHARING; STATIONS ABOARD SHIPS OR AIRCRAFT (14 QUESTIONS)

E1B - STATION RESTRICTIONS AND SPECIAL OPERATIONS: RESTRICTIONS ON STATION LOCATION; GENERAL OPERATING RESTRICTIONS, SPURIOUS EMISSIONS, CONTROL OPERATOR REIMBURSEMENT; ANTENNA STRUCTURE RESTRICTIONS; RACES OPERATIONS; NATIONAL QUIET ZONE (11 QUESTIONS)

E1C - DEFINITIONS AND RESTRICTIONS PERTAINING TO LOCAL, AUTOMATIC AND REMOTE CONTROL OPERATION; CONTROL OPERATOR RESPONSIBILITIES FOR REMOTE AND AUTOMATICALLY CONTROLLED STATIONS; IARP AND CEPT LICENSES; THIRD PARTY COMMUNICATIONS OVER AUTOMATICALLY CONTROLLED STATIONS (13 QUESTIONS)

E1D - AMATEUR SATELLITES: DEFINITIONS AND PURPOSE; LICENSE REQUIREMENTS FOR SPACE STATIONS; AVAILABLE FREQUENCIES AND BANDS; TELECOMMAND AND TELEMETRY OPERATIONS; RESTRICTIONS, AND SPECIAL PROVISIONS; NOTIFICATION REQUIREMENTS (11 QUESTIONS)

E1E - VOLUNTEER EXAMINER PROGRAM: DEFINITIONS; QUALIFICATIONS; PREPARATION AND ADMINISTRATION OF EXAMS; ACCREDITATION; QUESTION POOLS; DOCUMENTAITON REQUIRED (14 QUESTIONS)

E1F - MISCELLANEOUS RULES: EXTERNAL RF POWER AMPLIFIERS; BUSINESS COMMUNICATIONS; COMPENSATED COMMUNICATIONS; SPREAD SPECTRUM; AUXILIARY STATIONS; RECIPROCAL OPERATING PRIVILEGES; SPECIAL TEMPORARY AUTHORITY (12 QUESTIONS)

Subelement E2 - Operating Procedures [5 Groups]

E2A - AMATEUR RADIO IN SPACE: AMATEUR SATELLITES; ORBITAL MECHANICS; FREQUENCIES AND MODES; SATELLITE HARDWARE; SATELLITE OPERATIONS; EXPERIMENTAL TELEMETRY APPLICATIONS **(14 QUESTIONS)**

E2B - TELEVISION PRACTICES: FAST SCAN TELEVISION STANDARDS AND TECHNIQUES; SLOW SCAN TELEVISION STANDARDS AND TECHNIQUES **(19 QUESTIONS)**

E2C - OPERATING METHODS: CONTEST AND DX OPERATING; REMOTE OPERATION TECHNIQUES; CABRILLO FORMAT; QSLING; RF NETWORK CONNECTED SYSTEMS **(13 QUESTIONS)**

E2D - OPERATING METHODS: VHF AND UHF DIGITAL MODES AND PROCEDURES; APRS; EME PROCEDURES, METEOR SCATTER PROCEDURES **(14 QUESTIONS)**

E2E - OPERATING METHODS: OPERATING HF DIGITAL MODES **(13 QUESTIONS)**

Subelement E3 - Radio Wave Propagation [3 Groups]

E3A - ELECTROMAGNETIC WAVES; EARTH-MOON-EARTH COMMUNICATIONS; METEOR SCATTER; MICROWAVE TROPOSPHERIC AND SCATTER PROPAGATION; AURORA PROPAGATION **(17 QUESTIONS)**

E3B - TRANSEQUATORIAL PROPAGATION; LONG PATH; GRAY-LINE; MULTI-PATH; ORDINARY AND EXTRAORDINARY WAVES; CHORDAL HOP, SPORADIC E MECHANISMS **(14 QUESTIONS)**

E3C - RADIO-PATH HORIZON; LESS COMMON PROPAGATION MODES; PROPAGATION PREDICTION TECHNIQUES AND MODELING; SPACE WEATHER PARAMETERS AND AMATEUR RADIO **(15 QUESTIONS)**

Subelement E4 - Amateur Practices [5 Groups]

E4A- TEST EQUIPMENT: ANALOG AND DIGITAL INSTRUMENTS; SPECTRUM AND NETWORK ANALYZERS, ANTENNA ANALYZERS; OSCILLOSCOPES; RF MEASUREMENTS; COMPUTER AIDED MEASUREMENTS **(15 QUESTIONS)**

E4B - MEASUREMENT TECHNIQUE AND LIMITATIONS: INSTRUMENT ACCURACY AND PERFORMANCE LIMITATIONS; PROBES; TECHNIQUES TO MINIMIZE ERRORS; MEASUREMENT OF "Q"; INSTRUMENT CALIBRATION; S PARAMETERS; VECTOR NETWORK ANALYZERS **(17 QUESTIONS)**

E4C - RECEIVER PERFORMANCE CHARACTERISTICS, PHASE NOISE, NOISE FLOOR, IMAGE REJECTION, MDS, SIGNAL-TO-NOISE-RATIO; SELECTIVITY; EFFECTS OF SDR RECEIVER NON-LINEARITY **(17 QUESTIONS)**

E4D - RECEIVER PERFORMANCE CHARACTERISTICS: BLOCKING DYNAMIC RANGE; INTERMODULATION AND CROSS-MODULATION INTERFERENCE; 3RD ORDER INTERCEPT; DESENSITIZATION; PRESELECTOR **(14 QUESTIONS)**

E4E - NOISE SUPPRESSION: SYSTEM NOISE; ELECTRICAL APPLIANCE NOISE; LINE NOISE; LOCATING NOISE SOURCES; DSP NOISE REDUCTION; NOISE BLANKERS; GROUNDING FOR SIGNALS **(16 QUESTIONS)**

Subelement E5 - Electrical Principles [4 Groups]

E5A - RESONANCE AND Q: CHARACTERISTICS OF RESONANT CIRCUITS: SERIES AND PARALLEL RESONANCE; DEFINITIONS AND EFFECTS OF Q; HALF-POWER BANDWIDTH; PHASE RELATIONSHIPS IN REACTIVE CIRCUITS **(17 QUESTIONS)**

E5B - TIME CONSTANTS AND PHASE RELATIONSHIPS: RLC TIME CONSTANTS; DEFINITION; TIME CONSTANTS IN RL AND RC CIRCUITS; PHASE ANGLE BETWEEN VOLTAGE AND CURRENT; PHASE ANGLES OF SERIES RLC; PHASE ANGLE OF INDUCTANCE VS SUSCEPTANCE; ADMITTANCE AND SUSCEPTANCE **(13 QUESTIONS)**

E5C - COORDINATE SYSTEMS AND PHASORS IN ELECTRONICS: RECTANGULAR COORDINATES; POLAR COORDINATES; PHASORS **(17 QUESTIONS)**

E5D - AC AND RF ENERGY IN REAL CIRCUITS: SKIN EFFECT; ELECTROSTATIC AND ELECTROMAGNETIC FIELDS; REACTIVE POWER; POWER FACTOR; ELECTRICAL LENGTH OF CONDUCTORS AT UHF AND MICROWAVE FREQUENCIES **(18 QUESTIONS)**

Subelement E6 - Circuit Components [6 Groups]

E6A - SEMICONDUCTOR MATERIALS AND DEVICES: SEMICONDUCTOR MATERIALS; GERMANIUM, SILICON, P-TYPE, N-TYPE; TRANSISTOR TYPES: NPN, PNP, JUNCTION, FIELD-EFFECT TRANSISTORS: ENHANCEMENT MODE; DEPLETION MODE; MOS; CMOS; N-CHANNEL; P-CHANNEL **(17 QUESTIONS)**

E6B – DIODES **(13 QUESTIONS)**

E6C DIGITAL ICS: FAMILIES OF DIGITAL ICS; GATES; PROGRAMMABLE LOGIC DEVICES (PLDS) **(14 QUESTIONS)**

E6D - TOROIDAL AND SOLENOIDAL INDUCTORS: PERMEABILITY, CORE MATERIAL, SELECTING, WINDING; TRANSFORMERS; PIEZOELECTRIC DEVICES **(17 QUESTIONS)**

E6E - ANALOG ICS: MMICS, CCDS, DEVICE PACKAGES **(12 QUESTIONS)**

E6F - OPTICAL COMPONENTS: PHOTOCONDUCTIVE PRINCIPLES AND EFFECTS, PHOTOVOLTAIC SYSTEMS, OPTICAL COUPLERS, OPTICAL SENSORS, AND OPTOISOLATORS; LCDS **(14 QUESTIONS)**

Subelement E7 - Practical Circuits [8 Groups]

E7A - DIGITAL CIRCUITS: DIGITAL CIRCUIT PRINCIPLES AND LOGIC CIRCUITS: CLASSES OF LOGIC ELEMENTS; POSITIVE AND NEGATIVE LOGIC; FREQUENCY DIVIDERS; TRUTH TABLES **(12 QUESTIONS)**

E7B - AMPLIFIERS: CLASS OF OPERATION; VACUUM TUBE AND SOLID-STATE CIRCUITS; DISTORTION AND INTERMODULATION; SPURIOUS AND PARASITIC SUPPRESSION; MICROWAVE AMPLIFIERS; SWITCHING-TYPE AMPLIFIERS **(18 QUESTIONS)**

E7C - FILTERS AND MATCHING NETWORKS: TYPES OF NETWORKS; TYPES OF FILTERS; FILTER APPLICATIONS; FILTER CHARACTERISTICS; IMPEDANCE MATCHING; DSP FILTERING **(15 QUESTIONS)**

E7D - POWER SUPPLIES AND VOLTAGE REGULATORS; SOLAR ARRAY CHARGE CONTROLLERS **(16 QUESTIONS)**

E7E - MODULATION AND DEMODULATION: REACTANCE, PHASE AND BALANCED MODULATORS; DETECTORS; MIXER STAGES **(12 QUESTIONS)**

E7F - DSP FILTERING AND OTHER OPERATIONS; SOFTWARE DEFINED RADIO FUNDAMENTALS; DSP MODULATION AND DEMODULATION **(17 QUESTIONS)**

E7G - ACTIVE FILTERS AND OP-AMP CIRCUITS: ACTIVE AUDIO FILTERS; CHARACTERISTICS; BASIC CIRCUIT DESIGN; OPERATIONAL AMPLIFIERS **(12 QUESTIONS)**

E7H - OSCILLATORS AND SIGNAL SOURCES: TYPES OF OSCILLATORS; SYNTHESIZERS AND PHASE-LOCKED LOOPS; DIRECT DIGITAL SYNTHESIZERS; STABILIZING THERMAL DRIFT; MICROPHONICS; HIGH ACCURACY OSCILLATORS **(15 QUESTIONS)**

Subelement E8 - Signals and Emissions [4 Groups]

E8A - AC WAVEFORMS: SINE, SQUARE, SAWTOOTH AND IRREGULAR WAVEFORMS; AC MEASUREMENTS; AVERAGE AND PEP OF RF SIGNALS; FOURIER ANALYSIS; ANALOG TO DIGITAL CONVERSION: DIGITAL TO ANALOG CONVERSION **(13 QUESTIONS)**

E8B - MODULATION AND DEMODULATION: MODULATION METHODS; MODULATION INDEX AND DEVIATION RATIO; FREQUENCY AND TIME DIVISION MULTIPLEXING; ORTHOGONAL FREQUENCY DIVISION MULTIPLEXING **(11 QUESTIONS)**

E8C - DIGITAL SIGNALS: DIGITAL COMMUNICATION MODES; INFORMATION RATE VS BANDWIDTH; ERROR CORRECTION **(11 QUESTIONS)**

E8D - KEYING DEFECTS AND OVERMODULATION OF DIGITAL SIGNALS; DIGITAL CODES; SPREAD SPECTRUM **(12 QUESTIONS)**

Subelement E9 - Antennas and Transmission Lines [8 Groups]

E9A - BASIC ANTENNA PARAMETERS: RADIATION RESISTANCE, GAIN, BEAMWIDTH, EFFICIENCY, BEAMWIDTH; EFFECTIVE RADIATED POWER, POLARIZATION **(18 QUESTIONS)**

E9B - ANTENNA PATTERNS: E AND H PLANE PATTERNS; GAIN AS A FUNCTION OF PATTERN; ANTENNA DESIGN **(16 QUESTIONS)**

E9C - WIRE AND PHASED ARRAY ANTENNAS: RHOMBIC ANTENNAS; EFFECTS OF GROUND REFLECTIONS; E-OFF ANGLES; PRACTICAL WIRE ANTENNAS: ZEPPS, OCFD, LOOPS **(15 QUESTIONS)**

E9D - DIRECTIONAL ANTENNAS: GAIN; YAGI ANTENNAS; LOSSES; SWR BANDWIDTH; ANTENNA EFFICIENCY; SHORTENED AND MOBILE ANTENNAS; RF GROUNDING **(13 QUESTIONS)**

E9E - MATCHING: MATCHING ANTENNAS TO FEED LINES; PHASING LINES; POWER DIVIDERS **(13 QUESTIONS)**

E9F - TRANSMISSION LINES: CHARACTERISTICS OF OPEN AND SHORTED FEED LINES; 1/8 WAVELENGTH; 1/4 WAVELENGTH; 1/2 WAVELENGTH; FEED LINES: COAX VERSUS OPEN-WIRE; VELOCITY FACTOR; ELECTRICAL LENGTH; COAXIAL CABLE DIELECTRICS; VELOCITY FACTOR **(16 QUESTIONS)**

E9G - THE SMITH CHART **(11 QUESTIONS)**

E9H - RECEIVING ANTENNAS: RADIO DIRECTION FINDING ANTENNAS; BEVERAGE ANTENNAS; SPECIALIZED RECEIVING ANTENNAS; LONGWIRE RECEIVING ANTENNAS **(11 QUESTIONS)**

Subelement E0 – Safety - [1 Group]

<u>E0A - SAFETY: AMATEUR RADIO SAFETY PRACTICES; RF RADIATION HAZARDS; HAZARDOUS MATERIALS; GROUNDING</u> **(11 QUESTIONS)**

Main Study Effort

Now that you have learned how the Question Pool is organized by subelement and group, we get into the meat of the matter—the questions. The Extra Class Question Pool (valid 2016 to 2020) contains about 700 multiple choice questions. Each question is preceded by a code and numbering system, which is explained below. Also included here are study hints to help you focus your effort. Finally, also in this section, are the official graphics needed to answer some of the questions.

Studying the Question Pool in detail is very important. Exam questions and their multiple-choice answers are lifted VERBATIM from the Question Pool. Usually not even the sequence of the multiple choices is changed, i.e., what Answer A is in the Question Pool is usually what Answer A will look like in your exam. The same goes for B, C and D. What you study at home is most likely EXACTLY WORD FOR WORD what you will see on exam day. So let's talk about how each question is organized.

Question Pool Code and Numbering System

Each question will be followed by four multiple choice answers (A to D). The first line of every question in the Question Pool begins with the letter "E" for "Extra Class" followed by letters and numbers. Here is one example:

E1A09 (A) [97.219]
What is the first action you should take if your digital message forwarding station inadvertently forwards a communication that violates FCC rules?
A. Discontinue forwarding the communication as soon as you become aware of it
B. Notify the originating station that the communication does not comply with FCC rules
C. Notify the nearest FCC Field Engineer's office
D. Discontinue forwarding all messages

In this example the first line is "E1A09 (A) [97.219]." This refers to the question number, the right answer and any relevant regulation. It provides a short hand for easy cross reference when studying for the exam. But obviously, it will not be included with the questions on exam day.

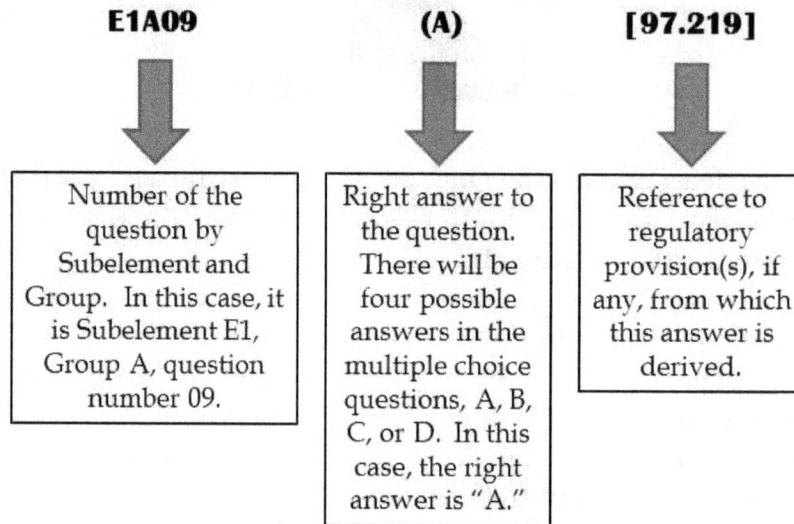

E1A09	**(A)**	**[97.219]**
Number of the question by Subelement and Group. In this case, it is Subelement E1, Group A, question number 09.	Right answer to the question. There will be four possible answers in the multiple choice questions, A, B, C, or D. In this case, the right answer is "A."	Reference to regulatory provision(s), if any, from which this answer is derived.

Study Hints

This main effort is just that, the section where you should devote most of your study time. Here are some hints to guide you:

- **Remember to Play the Odds.** There are 50 questions on every exam, one question selected from each of the 50 Groups in the Question Pool. You only need to answer 37 out of the 50 questions correctly to pass. But while each Group will have one question on the exam, some of the groups have only 11 questions while others have up to 19! But remember that you will see only one question from each Group, regardless of its size. Also, some group topics will be more challenging for you than others. More questions require more study time as do the more challenging topics. So do some strategizing and focus on what you can learn rather than spend extra time struggling with the rest (at least for the exam—certainly you will have a lifetime to learn all there is to know about Amateur Radio after you get your Extra Class License).
- **Study the Whole Multiple Choice Question.** Spend some time to familiarize yourself not only with the question and right answer, but also learn to recognize

the wrong answers as well. Because each multiple choice question will be lifted VERBATIM from the Question Pool for the exam, and most likely not even the sequence of the answers will change, this obviously works to your advantage. You will learn to spot the right answer more quickly. Also, in the event that several questions have similar answers (and there are more than a few of those) seeing the right answer positioned alongside all the wrong choices might just jog your memory.

- **<u>Take Bite-Size Pieces.</u>** It might take you several hours or more to review all 700 questions. Unless you have the time and patience, there is an easier way. Try instead at any single sitting to focus on one or two of the subelements or several of the groups. Whenever there is some down time, just grab this manual for some "Quick Study." Ten minutes here, twenty minutes there is lot more manageable in a busy day than trying to find several hours of quiet time all at once.

- **<u>Eliminate the Wrong Answers.</u>** One way to familiarize yourself with the whole multiple choice question is to see if you can figure out the right answer by eliminating all the wrong choices. Sometimes that is not possible, but sometimes the wrong choices are glaringly wrong and easy to eliminate, so give it a try.

- **<u>Spot Test with Online Practice Exams.</u>** Once you get some studying under your belt, you might want to test what you have learned by trying some of the online practice exams (just search "Amateur Radio Practice Exams"). But please don't rely on these tests as the primary method of study because they are a very inefficient way to learn. The tests randomly select 50 questions from the over 700 in the Question Pool. You could take multiple tests but the odds of seeing all the questions even once are very, very low. You can spend many hours taking many online tests and see some of the questions repeatedly, while chances are that other questions in the Question Pool you will not see at all. Tough luck if the first time you see a question is on exam day! So use the online tests, but only as a secondary method of study.

- **<u>Stick with It.</u>** The more you study, the more you learn, so hang in there!

Here are the graphics needed to answer some of the questions in sections E5, E6, E7, and E9 on the exam. Questions referring to graphics are a perennial favorite and therefore at least several are likely to appear again on exam day.

Figure E5-2

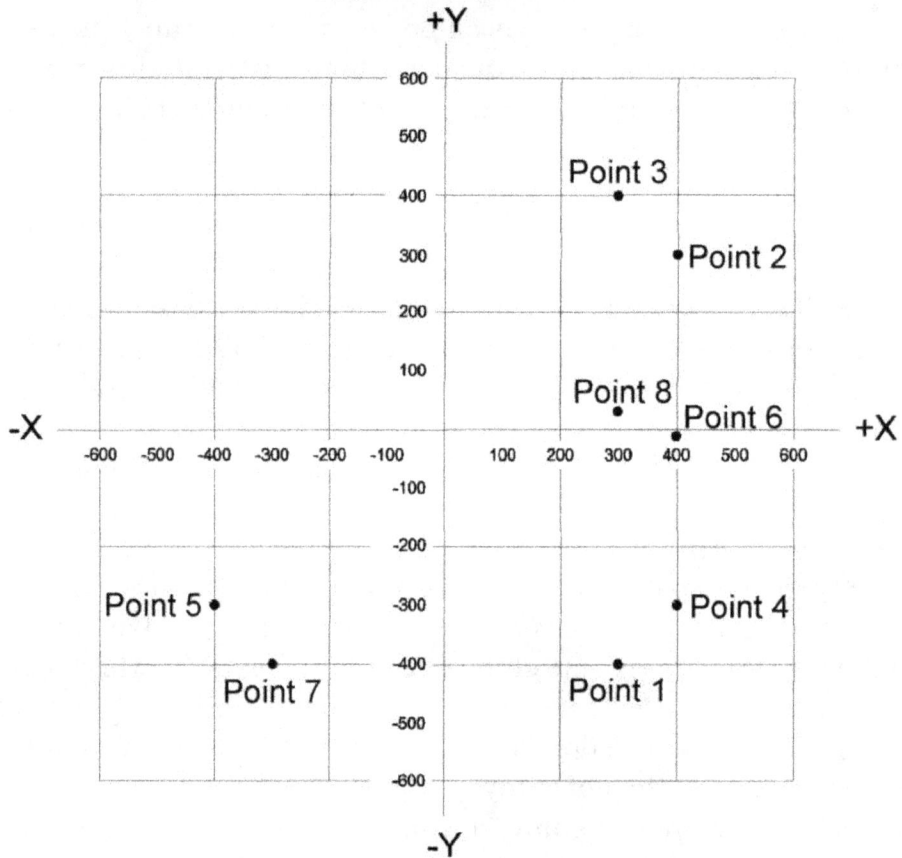

Figure E6-1

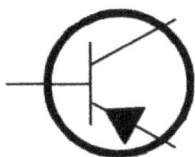

1

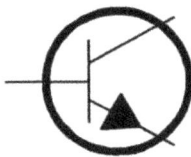

2

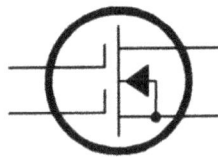

3

4

5

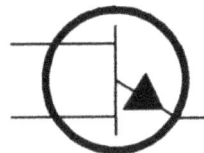

6

Figure E6-2

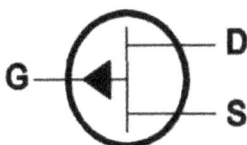

1

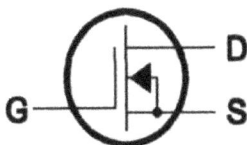

2

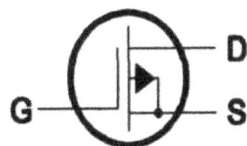

3

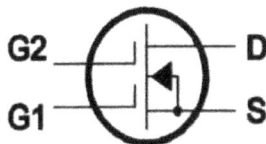

4

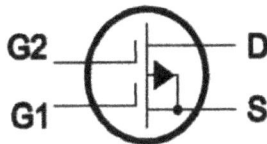

5

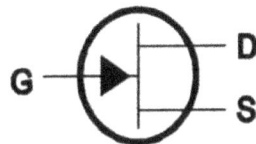

6

Figure E6-3

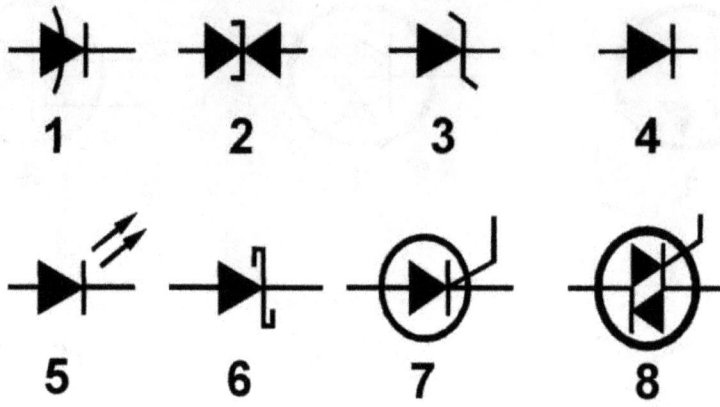

1 **2** **3** **4**

5 **6** **7** **8**

Figure E6-5

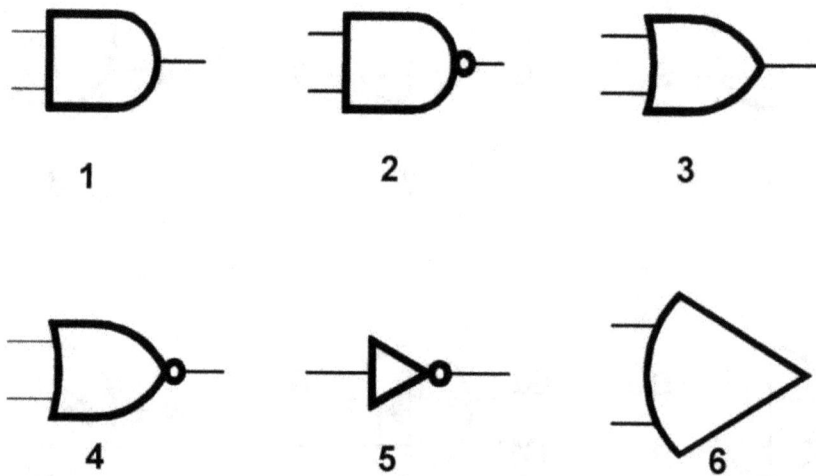

1 **2** **3**

4 **5** **6**

24

Figure E7-1

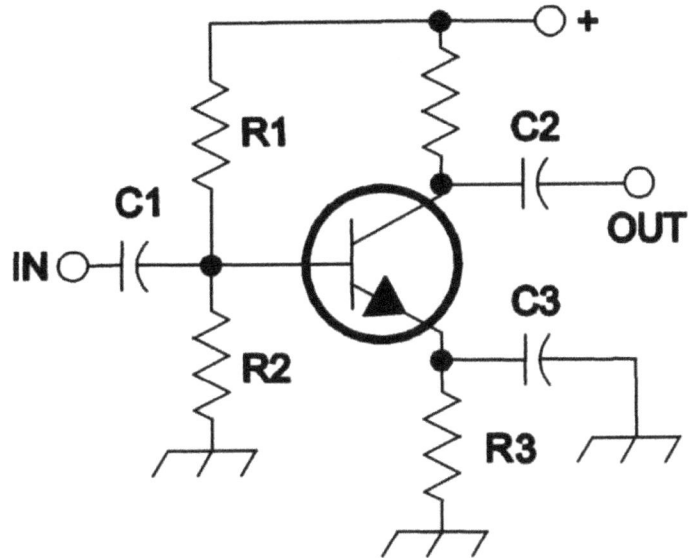

Figure E7-2

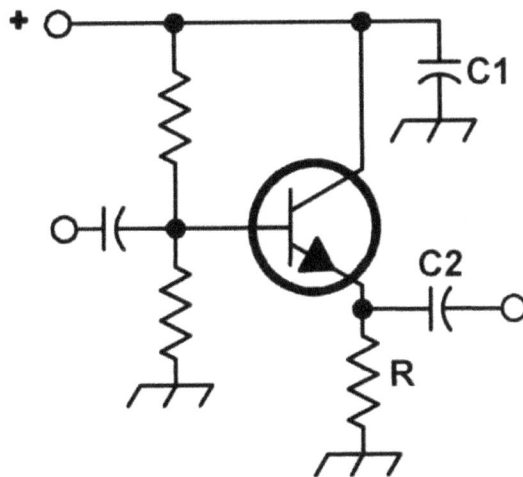

Figure E7- 3

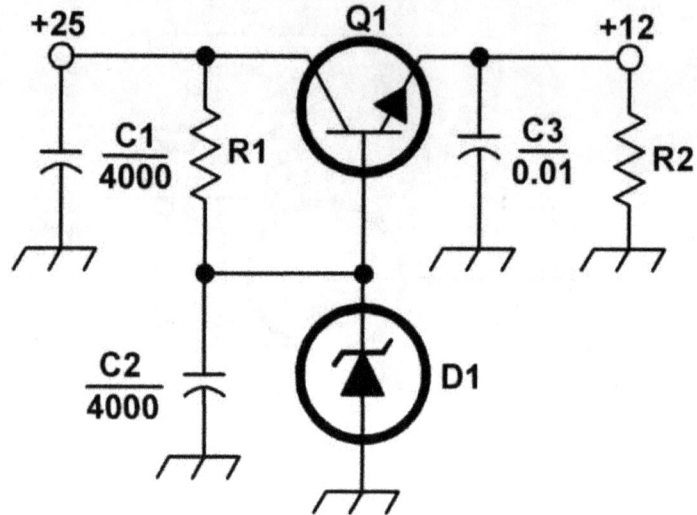

+25 Q1 +12

C1
4000 R1 C3
0.01 R2

C2
4000 D1

Figure E7-4

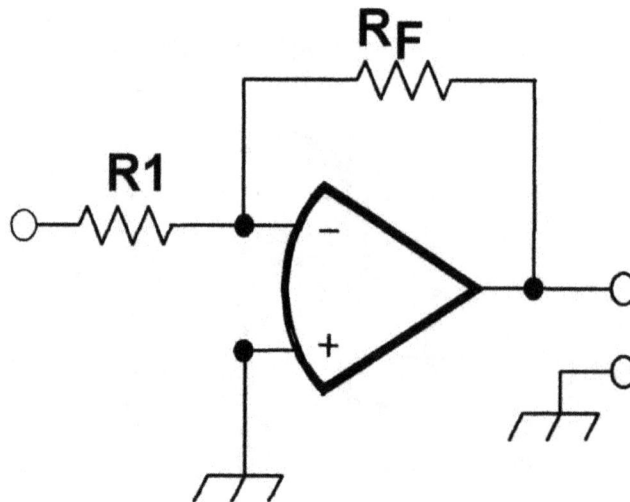

R_F

R1

Figure E9-1

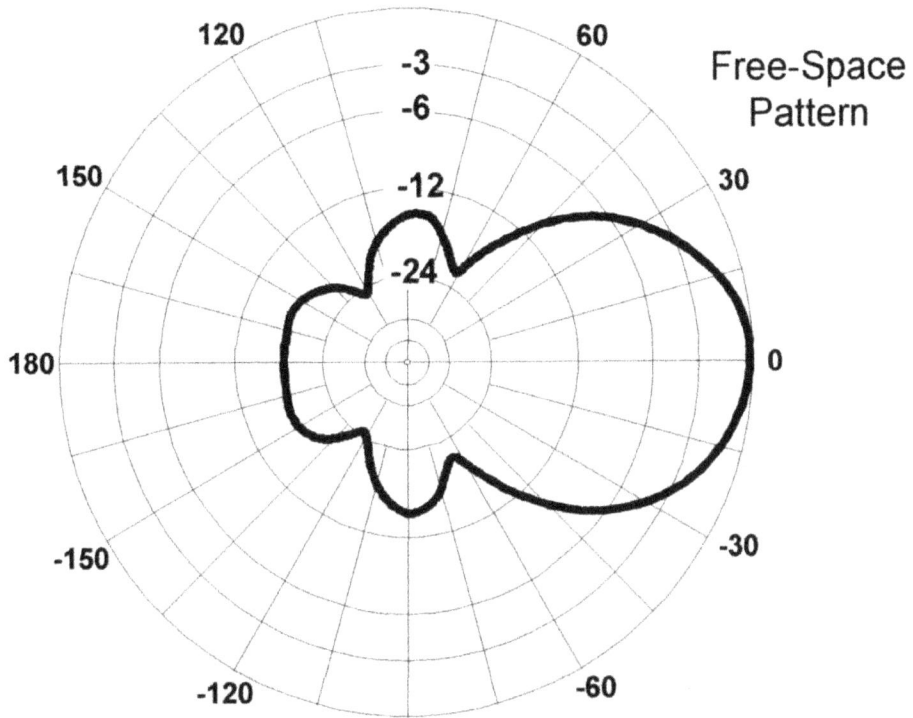

Free-Space Pattern

Figure E9-2

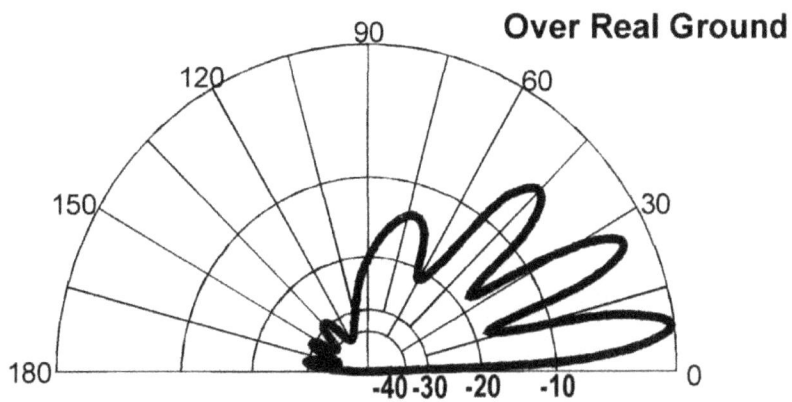

Over Real Ground

Figure E9-3

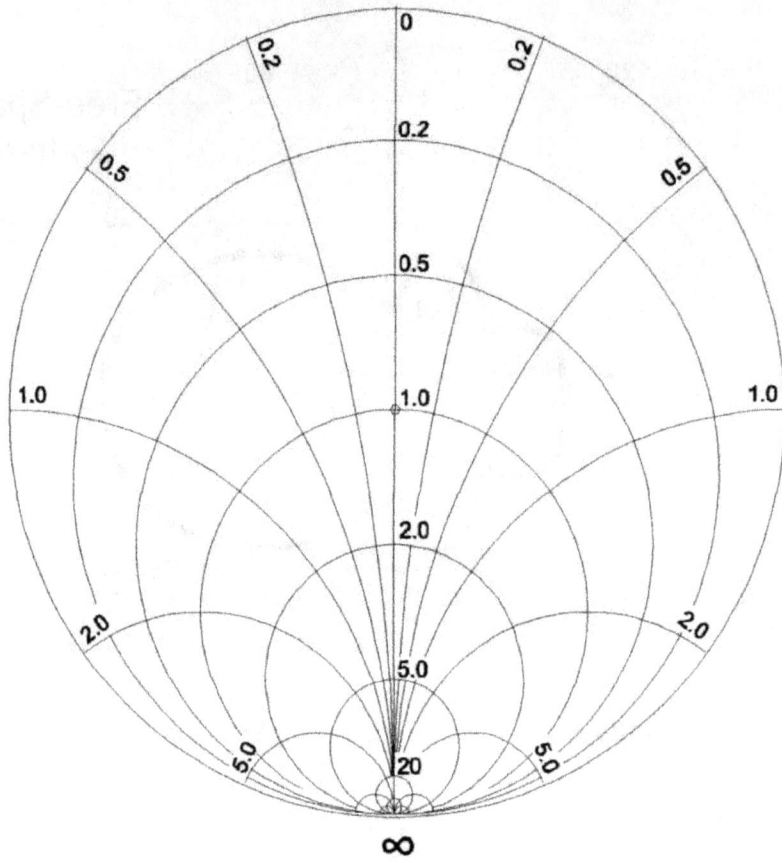

Extra Class Question Pool Valid 2016-2020

Subelement E1 – Commission's Rules [6 Groups]

E1A - OPERATING STANDARDS: FREQUENCY PRIVILEGES; EMISSION STANDARDS; AUTOMATIC MESSAGE FORWARDING; FREQUENCY SHARING; STATIONS ABOARD SHIPS OR AIRCRAFT

E1A01 (D) [97.301, 97.305]
When using a transceiver that displays the carrier frequency of phone signals, which of the following displayed frequencies represents the highest frequency at which a properly adjusted USB emission will be totally within the band?
A. The exact upper band edge
B. 300 Hz below the upper band edge
C. 1 kHz below the upper band edge
D. 3 kHz below the upper band edge

E1A02 (D) [97.301, 97.305]
When using a transceiver that displays the carrier frequency of phone signals, which of the following displayed frequencies represents the lowest frequency at which a properly adjusted LSB emission will be totally within the band?
A. The exact lower band edge
B. 300 Hz above the lower band edge
C. 1 kHz above the lower band edge
D. 3 kHz above the lower band edge

E1A03 (C) [97.301, 97.305]
With your transceiver displaying the carrier frequency of phone signals, you hear a station calling CQ on 14.349 MHz USB. Is it legal to return the call using upper sideband on the same frequency?
A. Yes, because you were not the station calling CQ
B. Yes, because the displayed frequency is within the 20 meter band
C. No, the sideband will extend beyond the band edge
D. No, U.S. stations are not permitted to use phone emissions above 14.340 MHz

E1A04 (C) [97.301, 97.305]
With your transceiver displaying the carrier frequency of phone signals, you hear a DX station calling CQ on 3.601 MHz LSB. Is it legal to return the call using lower sideband on the same frequency?
A. Yes, because the DX station initiated the contact
B. Yes, because the displayed frequency is within the 75 meter phone band segment
C. No, the sideband will extend beyond the edge of the phone band segment
D. No, U.S. stations are not permitted to use phone emissions below 3.610 MHz

E1A05 (C) [97.313]
What is the maximum power output permitted on the 60 meter band?
A. 50 watts PEP effective radiated power relative to an isotropic radiator
B. 50 watts PEP effective radiated power relative to a dipole
C. 100 watts PEP effective radiated power relative to the gain of a half-wave dipole
D. 100 watts PEP effective radiated power relative to an isotropic radiator

E1A06 (B) [97.15]
Where must the carrier frequency of a CW signal be set to comply with FCC rules for 60 meter operation?
A. At the lowest frequency of the channel
B. At the center frequency of the channel
C. At the highest frequency of the channel
D. On any frequency where the signal's sidebands are within the channel

E1A07 (D) [97.303]
Which amateur band requires transmission on specific channels rather than on a range of frequencies?
A. 12 meter band
B. 17 meter band
C. 30 meter band
D. 60 meter band

E1A08 (B) [97.219]
If a station in a message forwarding system inadvertently forwards a message that is in violation of FCC rules, who is primarily accountable for the rules violation?
A. The control operator of the packet bulletin board station
B. The control operator of the originating station
C. The control operators of all the stations in the system
D. The control operators of all the stations in the system not authenticating the source from which they accept communications

E1A09 (A) [97.219]
What is the first action you should take if your digital message forwarding station inadvertently forwards a communication that violates FCC rules?
A. Discontinue forwarding the communication as soon as you become aware of it
B. Notify the originating station that the communication does not comply with FCC rules
C. Notify the nearest FCC Field Engineer's office
D. Discontinue forwarding all messages

E1A10 (A) [97.11]
If an amateur station is installed aboard a ship or aircraft, what condition must be met before the station is operated?
A. Its operation must be approved by the master of the ship or the pilot in command of the aircraft
B. The amateur station operator must agree not to transmit when the main radio of the ship or aircraft is in use
C. The amateur station must have a power supply that is completely independent of the main ship or aircraft power supply
D. The amateur operator must have an FCC Marine or Aircraft endorsement on his or her amateur license

E1A11 (B) [97.5]
Which of the following describes authorization or licensing required when operating an amateur station aboard a U.S.-registered vessel in international waters?
A. Any amateur license with an FCC Marine or Aircraft endorsement
B. Any FCC-issued amateur license
C. Only General class or higher amateur licenses
D. An unrestricted Radiotelephone Operator Permit

E1A12 (C) [97.301, 97.305]
With your transceiver displaying the carrier frequency of CW signals, you hear a DX station's CQ on 3.500 MHz. Is it legal to return the call using CW on the same frequency?
A. Yes, the DX station initiated the contact
B. Yes, the displayed frequency is within the 80 meter CW band segment
C. No, one of the sidebands of the CW signal will be out of the band
D. No, U.S. stations are not permitted to use CW emissions below 3.525 MHz

E1A13 (B) [97.5]
Who must be in physical control of the station apparatus of an amateur station aboard any vessel or craft that is documented or registered in the United States?
A. Only a person with an FCC Marine Radio
B. Any person holding an FCC issued amateur license or who is authorized for alien reciprocal operation
C. Only a person named in an amateur station license grant
D. Any person named in an amateur station license grant or a person holding an unrestricted Radiotelephone Operator Permit

E1A14 (D) [97.303]
What is the maximum bandwidth for a data emission on 60 meters?
A. 60 Hz
B. 170 Hz
C. 1.5 kHz
D. 2.8 kHz

E1B - STATION RESTRICTIONS AND SPECIAL OPERATIONS: RESTRICTIONS ON STATION LOCATION; GENERAL OPERATING RESTRICTIONS, SPURIOUS EMISSIONS, CONTROL OPERATOR REIMBURSEMENT; ANTENNA STRUCTURE RESTRICTIONS; RACES OPERATIONS; NATIONAL QUIET ZONE

E1B01 (D) [97.3]
Which of the following constitutes a spurious emission?
A. An amateur station transmission made at random without the proper call sign identification
B. A signal transmitted to prevent its detection by any station other than the intended recipient
C. Any transmitted signal that unintentionally interferes with another licensed radio station
D. An emission outside its necessary bandwidth that can be reduced or eliminated without affecting the information transmitted

E1B02 (D) [97.13]
Which of the following factors might cause the physical location of an amateur station apparatus or antenna structure to be restricted?
A. The location is near an area of political conflict
B. The location is of geographical or horticultural importance
C. The location is in an ITU Zone designated for coordination with one or more foreign governments
D. The location is of environmental importance or significant in American history, architecture, or culture

E1B03 (A) [97.13]
Within what distance must an amateur station protect an FCC monitoring facility from harmful interference?
A. 1 mile
B. 3 miles
C. 10 miles
D. 30 miles

E1B04 (C) [97.13, 1.1305-1.1319]
What must be done before placing an amateur station within an officially designated wilderness area or wildlife preserve, or an area listed in the National Register of Historical Places?
A. A proposal must be submitted to the National Park Service
B. A letter of intent must be filed with the National Audubon Society
C. An Environmental Assessment must be submitted to the FCC
D. A form FSD-15 must be submitted to the Department of the Interior

E1B05 (C) [97.3]
What is the National Radio Quiet Zone?
A. An area in Puerto Rico surrounding the Arecibo Radio Telescope
B. An area in New Mexico surrounding the White Sands Test Area
C. An area surrounding the National Radio Astronomy Observatory
D. An area in Florida surrounding Cape Canaveral

E1B06 (A) [97.15]
Which of the following additional rules apply if you are installing an amateur station antenna at a site at or near a public use airport?
A. You may have to notify the Federal Aviation Administration and register it with the FCC as required by Part 17 of FCC rules
B. No special rules apply if your antenna structure will be less than 300 feet in height
C. You must file an Environmental Impact Statement with the EPA before construction begins
D. You must obtain a construction permit from the airport zoning authority

E1B07 (B) [97.307]
What is the highest modulation index permitted at the highest modulation frequency for angle modulation below 29.0 MHz?
A. 0.5
B. 1.0
C. 2.0
D. 3.0

E1B08 (D) [97.121]
What limitations may the FCC place on an amateur station if its signal causes interference to domestic broadcast reception, assuming that the receivers involved are of good engineering design?
A. The amateur station must cease operation
B. The amateur station must cease operation on all frequencies below 30 MHz
C. The amateur station must cease operation on all frequencies above 30 MHz
D. The amateur station must avoid transmitting during certain hours on frequencies that cause the interference

E1B09 (C) [97.407]
Which amateur stations may be operated under RACES rules?
A. Only those club stations licensed to Amateur Extra class operators
B. Any FCC-licensed amateur station except a Technician class
C. Any FCC-licensed amateur station certified by the responsible civil defense organization for the area served
D. Any FCC-licensed amateur station participating in the Military Auxiliary Radio System (MARS)

E1B10 (A) [97.407]
What frequencies are authorized to an amateur station operating under RACES rules?
A. All amateur service frequencies authorized to the control operator
B. Specific segments in the amateur service MF, HF, VHF and UHF bands
C. Specific local government channels
D. Military Auxiliary Radio System (MARS) channels

E1B11 (A) [97.307]
What is the permitted mean power of any spurious emission relative to the mean power of the fundamental emission from a station transmitter or external RF amplifier installed after January 1, 2003 and transmitting on a frequency below 30 MHZ?
A. At least 43 dB below
B. At least 53 dB below
C. At least 63 dB below
D. At least 73 dB below

E1C - DEFINITIONS AND RESTRICTIONS PERTAINING TO LOCAL, AUTOMATIC AND REMOTE CONTROL OPERATION; CONTROL OPERATOR RESPONSIBILITIES FOR REMOTE AND AUTOMATICALLY CONTROLLED STATIONS; IARP AND CEPT LICENSES; THIRD PARTY COMMUNICATIONS OVER AUTOMATICALLY CONTROLLED STATIONS

E1C01 (D) [97.3]
What is a remotely controlled station?
A. A station operated away from its regular home location
B. A station controlled by someone other than the licensee
C. A station operating under automatic control
D. A station controlled indirectly through a control link

E1C02 (A) [97.3, 97.109]
What is meant by automatic control of a station?
A. The use of devices and procedures for control so that the control operator does not have to be present at a control point
B. A station operating with its output power controlled automatically
C. Remotely controlling a station's antenna pattern through a directional control link
D. The use of a control link between a control point and a locally controlled station

E1C03 (B) [97.3, 97.109]
How do the control operator responsibilities of a station under automatic control differ from one under local control?
A. Under local control there is no control operator
B. Under automatic control the control operator is not required to be present at the control point
C. Under automatic control there is no control operator
D. Under local control a control operator is not required to be present at a control point

E1C04 (A)
What is meant by IARP?
A. An international amateur radio permit that allows U.S. amateurs to operate in certain countries of the Americas
B. The internal amateur radio practices policy of the FCC
C. An indication of increased antenna reflected power
D. A forecast of intermittent aurora radio propagation

E1C05 (A) [97.221(c)(1),[97.115(c)]
When may an automatically controlled station originate third party communications?
A. Never
B. Only when transmitting RTTY or data emissions
C. When agreed upon by the sending or receiving station
D. When approved by the National Telecommunication and Information Administration

E1C06 (C) [97.109]
Which of the following statements concerning remotely controlled amateur stations is true?
A. Only Extra Class operators may be the control operator of a remote station
B. A control operator need not be present at the control point
C. A control operator must be present at the control point
D. Repeater and auxiliary stations may not be remotely controlled

E1C07 (C) [97.3]
What is meant by local control?
A. Controlling a station through a local auxiliary link
B. Automatically manipulating local station controls
C. Direct manipulation of the transmitter by a control operator
D. Controlling a repeater using a portable handheld transceiver

E1C08 (B) [97.213]
What is the maximum permissible duration of a remotely controlled station's transmissions if its control link malfunctions?
A. 30 seconds
B. 3 minutes
C. 5 minutes
D. 10 minutes

E1C09 (D) [97.205]
Which of these ranges of frequencies is available for an automatically controlled repeater operating below 30 MHz?
A. 18.110 MHz - 18.168 MHz
B. 24.940 MHz - 24.990 MHz
C. 10.100 MHz - 10.150 MHz
D. 29.500 MHz - 29.700 MHz

E1C10 (B) [97.113]
What types of amateur stations may automatically retransmit the radio signals of other amateur stations?
A. Only beacon, repeater or space stations
B. Only auxiliary, repeater or space stations
C. Only earth stations, repeater stations or model craft
D. Only auxiliary, beacon or space stations

E1C11 (A) [97.5]
Which of the following operating arrangements allows an FCC-licensed U.S. citizen to operate in many European countries, and alien amateurs from many European countries to operate in the U.S.?
A. CEPT agreement
B. IARP agreement
C. ITU reciprocal license
D. All of these choices are correct

E1C12 (C) [97.117]
What types of communications may be transmitted to amateur stations in foreign countries?
A. Business-related messages for non-profit organizations
B. Messages intended for connection to users of the maritime satellite service
C. Communications incidental to the purpose of the amateur service and remarks of a personal nature
D. All of these choices are correct

E1C13 (C)
Which of the following is required in order to operate in accordance with CEPT rules in foreign countries where permitted?
A. You must identify in the official language of the country in which you are operating
B. The U.S. embassy must approve of your operation
C. You must bring a copy of FCC Public Notice DA 11-221
D. You must append "/CEPT" to your call sign

E1D - AMATEUR SATELLITES: DEFINITIONS AND PURPOSE; LICENSE REQUIREMENTS FOR SPACE STATIONS; AVAILABLE FREQUENCIES AND BANDS; TELECOMMAND AND TELEMETRY OPERATIONS; RESTRICTIONS, AND SPECIAL PROVISIONS; NOTIFICATION REQUIREMENTS

E1D01 (A) [97.3]
What is the definition of the term telemetry?
A. One-way transmission of measurements at a distance from the measuring instrument
B. Two-way radiotelephone transmissions in excess of 1000 feet
C. Two-way single channel transmissions of data
D. One-way transmission that initiates, modifies, or terminates the functions of a device at a distance

E1D02 (C) [97.3]
What is the amateur satellite service?
A. A radio navigation service using satellites for the purpose of self training, intercommunication and technical studies carried out by amateurs
B. A spacecraft launching service for amateur-built satellites
C. A radio communications service using amateur radio stations on satellites
D. A radio communications service using stations on Earth satellites for public service broadcast

E1D03 (B) [97.3]
What is a telecommand station in the amateur satellite service?
A. An amateur station located on the Earth's surface for communication with other Earth stations by means of Earth satellites
B. An amateur station that transmits communications to initiate, modify or terminate functions of a space station
C. An amateur station located more than 50 km above the Earth's surface
D. An amateur station that transmits telemetry consisting of measurements of upper atmosphere

E1D04 (A) [97.3]
What is an Earth station in the amateur satellite service?
A. An amateur station within 50 km of the Earth's surface intended for communications with amateur stations by means of objects in space
B. An amateur station that is not able to communicate using amateur satellites
C. An amateur station that transmits telemetry consisting of measurement of upper atmosphere
D. Any amateur station on the surface of the Earth

E1D05 (C) [97.207]
What class of licensee is authorized to be the control operator of a space station?
A. All except Technician Class
B. Only General, Advanced or Amateur Extra Class
C. Any class with appropriate operator privileges
D. Only Amateur Extra Class

E1D06 (A) [97.207]
Which of the following is a requirement of a space station?
A. The space station must be capable of terminating transmissions by telecommand when directed by the FCC
B. The space station must cease all transmissions after 5 years
C. The space station must be capable of changing its orbit whenever such a change is ordered by NASA
D. All of these choices are correct

E1D07 (A) [97.207]
Which amateur service HF bands have frequencies authorized for space stations?
A. Only the 40 m, 20 m, 17 m, 15 m, 12 m and 10 m bands
B. Only the 40 m, 20 m, 17 m, 15 m and 10 m bands
C. Only the 40 m, 30 m, 20 m, 15 m, 12 m and 10 m bands
D. All HF bands

E1D08 (D) [97.207]
Which VHF amateur service bands have frequencies available for space stations?
A. 6 meters and 2 meters
B. 6 meters, 2 meters, and 1.25 meters
C. 2 meters and 1.25 meters
D. 2 meters

E1D09 (B) [97.207]
Which UHF amateur service bands have frequencies available for a space station?
A. 70 cm only
B. 70 cm and 13 cm
C. 70 cm and 33 cm
D. 33 cm and 13 cm

E1D10 (B) [97.211]
Which amateur stations are eligible to be telecommand stations?
A. Any amateur station designated by NASA
B. Any amateur station so designated by the space station licensee, subject to the privileges of the class of operator license held by the control operator
C. Any amateur station so designated by the ITU
D. All of these choices are correct

E1D11 (D) [97.209]
Which amateur stations are eligible to operate as Earth stations?
A. Any amateur station whose licensee has filed a pre-space notification with the FCC's International Bureau
B. Only those of General, Advanced or Amateur Extra Class operators
C. Only those of Amateur Extra Class operators
D. Any amateur station, subject to the privileges of the class of operator license held by the control operator

E1E - VOLUNTEER EXAMINER PROGRAM: DEFINITIONS; QUALIFICATIONS; PREPARATION AND ADMINISTRATION OF EXAMS; ACCREDITATION; QUESTION POOLS; DOCUMENTATION REQUIREMENTS

E1E01 (D) [97.509]
What is the minimum number of qualified VEs required to administer an Element 4 amateur operator license examination?
A. 5
B. 2
C. 4
D. 3

E1E02 (C) [97.523]
Where are the questions for all written U.S. amateur license examinations listed?
A. In FCC Part 97
B. In a question pool maintained by the FCC
C. In a question pool maintained by all the VECs
D. In the appropriate FCC Report and Order

E1E03 (C) [97.521]
What is a Volunteer Examiner Coordinator?
A. A person who has volunteered to administer amateur operator license examinations
B. A person who has volunteered to prepare amateur operator license examinations
C. An organization that has entered into an agreement with the FCC to coordinate amateur operator license examinations
D. The person who has entered into an agreement with the FCC to be the VE session manager

E1E04 (D) [97.509, 97.525]
Which of the following best describes the Volunteer Examiner accreditation process?
A. Each General, Advanced and Amateur Extra Class operator is automatically accredited as a VE when the license is granted
B. The amateur operator applying must pass a VE examination administered by the FCC Enforcement Bureau
C. The prospective VE obtains accreditation from the FCC
D. The procedure by which a VEC confirms that the VE applicant meets FCC requirements to serve as an examiner

E1E05 (B) [97.503]
What is the minimum passing score on amateur operator license examinations?
A. Minimum passing score of 70%
B. Minimum passing score of 74%
C. Minimum passing score of 80%
D. Minimum passing score of 77%

E1E06 (C) [97.509]
Who is responsible for the proper conduct and necessary supervision during an amateur operator license examination session?
A. The VEC coordinating the session
B. The FCC
C. Each administering VE
D. The VE session manager

E1E07 (B) [97.509]
What should a VE do if a candidate fails to comply with the examiner's instructions during an amateur operator license examination?
A. Warn the candidate that continued failure to comply will result in termination of the examination
B. Immediately terminate the candidate's examination
C. Allow the candidate to complete the examination, but invalidate the results
D. Immediately terminate everyone's examination and close the session

E1E08 (C) [97.509]
To which of the following examinees may a VE not administer an examination?
A. Employees of the VE
B. Friends of the VE
C. Relatives of the VE as listed in the FCC rules
D. All of these choices are correct

E1E09 (A) [97.509]
What may be the penalty for a VE who fraudulently administers or certifies an examination?
A. Revocation of the VE's amateur station license grant and the suspension of the VE's amateur operator license grant
B. A fine of up to $1000 per occurrence
C. A sentence of up to one year in prison
D. All of these choices are correct

E1E10 (C) [97.509]
What must the administering VEs do after the administration of a successful examination for an amateur operator license?
A. They must collect and send the documents to the NCVEC for grading
B. They must collect and submit the documents to the coordinating VEC for grading
C. They must submit the application document to the coordinating VEC according to the coordinating VEC instructions
D. They must collect and send the documents to the FCC according to instructions

E1E11 (B) [97.509]
What must the VE team do if an examinee scores a passing grade on all examination elements needed for an upgrade or new license?
A. Photocopy all examination documents and forward them to the FCC for processing
B. Three VEs must certify that the examinee is qualified for the license grant and that they have complied with the administering VE requirements
C. Issue the examinee the new or upgrade license
D. All these choices are correct

E1E12 (A) [97.509]
What must the VE team do with the application form if the examinee does not pass the exam?
A. Return the application document to the examinee
B. Maintain the application form with the VEC's records
C. Send the application form to the FCC and inform the FCC of the grade
D. Destroy the application form

E1E13 (B) [97.509]
Which of these choices is an acceptable method for monitoring the applicants if a VEC opts to conduct an exam session remotely?
A. Record the exam session on video tape for later review by the VE team
B. Use a real-time video link and the Internet to connect the exam session to the observing VEs
C. The exam proctor observes the applicants and reports any violations
D. Have each applicant sign an affidavit stating that all session rules were followed

E1E14 (A) [97.527]
For which types of out-of-pocket expenses do the Part 97 rules state that VEs and VECs may be reimbursed?
A. Preparing, processing, administering and coordinating an examination for an amateur radio license
B. Teaching an amateur operator license examination preparation course
C. No expenses are authorized for reimbursement
D. Providing amateur operator license examination preparation training materials

E1F - MISCELLANEOUS RULES: EXTERNAL RF POWER AMPLIFIERS; BUSINESS COMMUNICATIONS; COMPENSATED COMMUNICATIONS; SPREAD SPECTRUM; AUXILIARY STATIONS; RECIPROCAL OPERATING PRIVILEGES; SPECIAL TEMPORARY AUTHORITY

E1F01 (B) [97.305]
On what frequencies are spread spectrum transmissions permitted?
A. Only on amateur frequencies above 50 MHz
B. Only on amateur frequencies above 222 MHz
C. Only on amateur frequencies above 420 MHz
D. Only on amateur frequencies above 144 MHz

E1F02 (C) [97.107]
What privileges are authorized in the U.S. to persons holding an amateur service license granted by the Government of Canada?
A. None, they must obtain a U.S. license
B. All privileges of the Extra Class license
C. The operating terms and conditions of the Canadian amateur service license, not to exceed U.S. Extra Class privileges
D. Full privileges, up to and including those of the Extra Class License, on the 80, 40, 20, 15, and 10 meter bands

E1F03 (A) [97.315]
Under what circumstances may a dealer sell an external RF power amplifier capable of operation below 144 MHz if it has not been granted FCC certification?
A. It was purchased in used condition from an amateur operator and is sold to another amateur operator for use at that operator's station
B. The equipment dealer assembled it from a kit
C. It was imported from a manufacturer in a country that does not require certification of RF power amplifiers
D. It was imported from a manufacturer in another country and was certificated by that country's government

E1F04 (A) [97.3]
Which of the following geographic descriptions approximately describes "Line A"?
A. A line roughly parallel to and south of the U.S.-Canadian border
B. A line roughly parallel to and west of the U.S. Atlantic coastline
C. A line roughly parallel to and north of the U.S.-Mexican border and Gulf coastline
D. A line roughly parallel to and east of the U.S. Pacific coastline

E1F05 (D) [97.303]
Amateur stations may not transmit in which of the following frequency segments if they are located in the contiguous 48 states and north of Line A?
A. 440 MHz - 450 MHz
B. 53 MHz - 54 MHz
C. 222 MHz - 223 MHz
D. 420 MHz - 430 MHz

E1F06 (A) [1.931]
Under what circumstances might the FCC issue a Special Temporary Authority (STA) to an amateur station?
A. To provide for experimental amateur communications
B. To allow regular operation on Land Mobile channels
C. To provide additional spectrum for personal use
D. To provide temporary operation while awaiting normal licensing

E1F07 (D) [97.113]
When may an amateur station send a message to a business?
A. When the total money involved does not exceed $25
B. When the control operator is employed by the FCC or another government agency
C. When transmitting international third-party communications
D. When neither the amateur nor his or her employer has a pecuniary interest in the communications

E1F08 (A) [97.113]
Which of the following types of amateur station communications are prohibited?
A. Communications transmitted for hire or material compensation, except as otherwise provided in the rules
B. Communications that have a political content, except as allowed by the Fairness Doctrine
C. Communications that have a religious content
D. Communications in a language other than English

E1F09 (D) [97.311]
Which of the following conditions apply when transmitting spread spectrum emission?
A. A station transmitting SS emission must not cause harmful interference to other stations employing other authorized emissions
B. The transmitting station must be in an area regulated by the FCC or in a country that permits SS emissions
C. The transmission must not be used to obscure the meaning of any communication
D. All of these choices are correct

E1F10 (C) [97.313]
What is the maximum permitted transmitter peak envelope power for an amateur station transmitting spread spectrum communications?
A. 1 W
B. 1.5 W
C. 10 W
D. 1.5 kW

E1F11 (D) [97.317]
Which of the following best describes one of the standards that must be met by an external RF power amplifier if it is to qualify for a grant of FCC certification?
A. It must produce full legal output when driven by not more than 5 watts of mean RF input power
B. It must be capable of external RF switching between its input and output networks
C. It must exhibit a gain of 0 dB or less over its full output range
D. It must satisfy the FCC's spurious emission standards when operated at the lesser of 1500 watts or its full output power

E1F12 (B) [97.201]
Who may be the control operator of an auxiliary station?
A. Any licensed amateur operator
B. Only Technician, General, Advanced or Amateur Extra Class operators
C. Only General, Advanced or Amateur Extra Class operators
D. Only Amateur Extra Class operators

Subelement E2 - Operating Procedures [5 Groups]

E2A - AMATEUR RADIO IN SPACE: AMATEUR SATELLITES; ORBITAL MECHANICS; FREQUENCIES AND MODES; SATELLITE HARDWARE; SATELLITE OPERATIONS; EXPERIMENTAL TELEMETRY APPLICATIONS

E2A01 (C)
What is the direction of an ascending pass for an amateur satellite?
A. From west to east
B. From east to west
C. From south to north
D. From north to south

E2A02 (A)
What is the direction of a descending pass for an amateur satellite?
A. From north to south
B. From west to east
C. From east to west
D. From south to north

E2A03 (C)
What is the orbital period of an Earth satellite?
A. The point of maximum height of a satellite's orbit
B. The point of minimum height of a satellite's orbit
C. The time it takes for a satellite to complete one revolution around the Earth
D. The time it takes for a satellite to travel from perigee to apogee

E2A04 (B)
What is meant by the term mode as applied to an amateur radio satellite?
A. The type of signals that can be relayed through the satellite
B. The satellite's uplink and downlink frequency bands
C. The satellite's orientation with respect to the Earth
D. Whether the satellite is in a polar or equatorial orbit

E2A05 (D)
What do the letters in a satellite's mode designator specify?
A. Power limits for uplink and downlink transmissions
B. The location of the ground control station
C. The polarization of uplink and downlink signals
D. The uplink and downlink frequency ranges

E2A06 (A)
On what band would a satellite receive signals if it were operating in mode U/V?
A. 435 MHz - 438 MHz
B. 144 MHz - 146 MHz
C. 50.0 MHz - 50.2 MHz
D. 29.5 MHz - 29.7 MHz

E2A07 (D)
Which of the following types of signals can be relayed through a linear transponder?
A. FM and CW
B. SSB and SSTV
C. PSK and Packet
D. All of these choices are correct

E2A08 (B)
Why should effective radiated power to a satellite which uses a linear transponder be limited?
A. To prevent creating errors in the satellite telemetry
B. To avoid reducing the downlink power to all other users
C. To prevent the satellite from emitting out-of-band signals
D. To avoid interfering with terrestrial QSOs

E2A09 (A)
What do the terms L band and S band specify with regard to satellite communications?
A. The 23 centimeter and 13 centimeter bands
B. The 2 meter and 70 centimeter bands
C. FM and Digital Store-and-Forward systems
D. Which sideband to use

E2A10 (A)
Why may the received signal from an amateur satellite exhibit a rapidly repeating fading effect?
A. Because the satellite is spinning
B. Because of ionospheric absorption
C. Because of the satellite's low orbital altitude
D. Because of the Doppler Effect

E2A11 (B)
What type of antenna can be used to minimize the effects of spin modulation and Faraday rotation?
A. A linearly polarized antenna
B. A circularly polarized antenna
C. An isotropic antenna
D. A log-periodic dipole array

E2A12 (D)
What is one way to predict the location of a satellite at a given time?
A. By means of the Doppler data for the specified satellite
B. By subtracting the mean anomaly from the orbital inclination
C. By adding the mean anomaly to the orbital inclination
D. By calculations using the Keplerian elements for the specified satellite

E2A13 (B)
What type of satellite appears to stay in one position in the sky?
A. HEO
B. Geostationary
C. Geomagnetic
D. LEO

E2A14 (C)
What technology is used to track, in real time, balloons carrying amateur radio transmitters?
A. Radar
B. Bandwidth compressed LORAN
C. APRS
D. Doppler shift of beacon signals

E2B - TELEVISION PRACTICES: FAST SCAN TELEVISION STANDARDS AND TECHNIQUES; SLOW SCAN TELEVISION STANDARDS AND TECHNIQUES

E2B01 (A)
How many times per second is a new frame transmitted in a fast-scan (NTSC) television system?
A. 30
B. 60
C. 90
D. 120

E2B02 (C)
How many horizontal lines make up a fast-scan (NTSC) television frame?
A. 30
B. 60
C. 525
D. 1080

E2B03 D)
How is an interlaced scanning pattern generated in a fast-scan (NTSC) television system?
A. By scanning two fields simultaneously
B. By scanning each field from bottom to top
C. By scanning lines from left to right in one field and right to left in the next
D. By scanning odd numbered lines in one field and even numbered lines in the next

E2B04 (B)
What is blanking in a video signal?
A. Synchronization of the horizontal and vertical sync pulses
B. Turning off the scanning beam while it is traveling from right to left or from bottom to top
C. Turning off the scanning beam at the conclusion of a transmission
D. Transmitting a black and white test pattern

E2B05 (C)
Which of the following is an advantage of using vestigial sideband for standard fast-scan TV transmissions?
A. The vestigial sideband carries the audio information
B. The vestigial sideband contains chroma information
C. Vestigial sideband reduces bandwidth while allowing for simple video detector circuitry
D. Vestigial sideband provides high frequency emphasis to sharpen the picture

E2B06 (A)
What is vestigial sideband modulation?
A. Amplitude modulation in which one complete sideband and a portion of the other are transmitted
B. A type of modulation in which one sideband is inverted
C. Narrow-band FM modulation achieved by filtering one sideband from the audio before frequency modulating the carrier
D. Spread spectrum modulation achieved by applying FM modulation following single sideband amplitude modulation

E2B07 (B)
What is the name of the signal component that carries color information in NTSC video?
A. Luminance
B. Chroma
C. Hue
D. Spectral Intensity

E2B08 (D)
Which of the following is a common method of transmitting accompanying audio with amateur fast-scan television?
A. Frequency-modulated sub-carrier
B. A separate VHF or UHF audio link
C. Frequency modulation of the video carrier
D. All of these choices are correct

E2B09 (D)
What hardware, other than a receiver with SSB capability and a suitable computer, is needed to decode SSTV using Digital Radio Mondiale (DRM)?
A. A special IF converter
B. A special front end limiter
C. A special notch filter to remove synchronization pulses
D. No other hardware is needed

E2B10 (A)
Which of the following is an acceptable bandwidth for Digital Radio Mondiale (DRM) based voice or SSTV digital transmissions made on the HF amateur bands?
A. 3 KHz
B. 10 KHz
C. 15 KHz
D. 20 KHz

E2B11 (B)
What is the function of the Vertical Interval Signaling (VIS) code sent as part of an SSTV transmission?
A. To lock the color burst oscillator in color SSTV images
B. To identify the SSTV mode being used
C. To provide vertical synchronization
D. To identify the call sign of the station transmitting

E2B12 (D)
How are analog SSTV images typically transmitted on the HF bands?
A. Video is converted to equivalent Baudot representation
B. Video is converted to equivalent ASCII representation
C. Varying tone frequencies representing the video are transmitted using PSK
D. Varying tone frequencies representing the video are transmitted using single sideband

E2B13 (C)
How many lines are commonly used in each frame of an amateur slow-scan color television picture?
A. 30 or 60
B. 60 or 100
C. 128 or 256
D. 180 or 360

E2B14 (A)
What aspect of an amateur slow-scan television signal encodes the brightness of the picture?
A. Tone frequency
B. Tone amplitude
C. Sync amplitude
D. Sync frequency

E2B15 (A)
What signals SSTV receiving equipment to begin a new picture line?
A. Specific tone frequencies
B. Elapsed time
C. Specific tone amplitudes
D. A two-tone signal

E2B16 (D)
Which is a video standard used by North American Fast Scan ATV stations?
A. PAL
B. DRM
C. Scottie
D. NTSC

E2B17 (B)
What is the approximate bandwidth of a slow-scan TV signal?
A. 600 Hz
B. 3 kHz
C. 2 MHz
D. 6 MHz

E2B18 (D)
On which of the following frequencies is one likely to find FM ATV transmissions?
A. 14.230 MHz
B. 29.6 MHz
C. 52.525 MHz
D. 1255 MHz

E2B19 (C)
What special operating frequency restrictions are imposed on slow scan TV transmissions?
A. None; they are allowed on all amateur frequencies
B. They are restricted to 7.245 MHz, 14.245 MHz, 21.345 MHz, and 28.945 MHz
C. They are restricted to phone band segments and their bandwidth can be no greater than that of a voice signal of the same modulation type
D. They are not permitted above 54 MHz

E2C - OPERATING METHODS: CONTEST AND DX OPERATING; REMOTE OPERATION TECHNIQUES; CABRILLO FORMAT; QSLING; RF NETWORK CONNECTED SYSTEMS

E2C01 (A)
Which of the following is true about contest operating?
A. Operators are permitted to make contacts even if they do not submit a log
B. Interference to other amateurs is unavoidable and therefore acceptable
C. It is mandatory to transmit the call sign of the station being worked as part of every transmission to that station
D. Every contest requires a signal report in the exchange

E2C02 (A)
Which of the following best describes the term self-spotting in regards to HF contest operating?
A. The generally prohibited practice of posting one's own call sign and frequency on a spotting network
B. The acceptable practice of manually posting the call signs of stations on a spotting network
C. A manual technique for rapidly zero beating or tuning to a station's frequency before calling that station
D. An automatic method for rapidly zero beating or tuning to a station's frequency before calling that station

E2C03 (A)
From which of the following bands is amateur radio contesting generally excluded?
A. 30 m
B. 6 m
C. 2 m
D. 33 cm

E2C04 (A)
What type of transmission is most often used for a ham radio mesh network?
A. Spread spectrum in the 2.4 GHz band
B. Multiple Frequency Shift Keying in the 10 GHz band
C. Store and forward on the 440 MHz band
D. Frequency division multiplex in the 24 GHz band

E2C05 (B)
What is the function of a DX QSL Manager?
A. To allocate frequencies for DXpeditions
B. To handle the receiving and sending of confirmation cards for a DX station
C. To run a net to allow many stations to contact a rare DX station
D. To relay calls to and from a DX station

E2C06 (C)
During a VHF/UHF contest, in which band segment would you expect to find the highest level of activity?
A. At the top of each band, usually in a segment reserved for contests
B. In the middle of each band, usually on the national calling frequency
C. In the weak signal segment of the band, with most of the activity near the calling frequency
D. In the middle of the band, usually 25 kHz above the national calling frequency

E2C07 (A)
What is the Cabrillo format?
A. A standard for submission of electronic contest logs
B. A method of exchanging information during a contest QSO
C. The most common set of contest rules
D. The rules of order for meetings between contest sponsors

E2C08 (B)
Which of the following contacts may be confirmed through the U.S. QSL bureau system?
A. Special event contacts between stations in the U.S.
B. Contacts between a U.S. station and a non-U.S. station
C. Repeater contacts between U.S. club members
D. Contacts using tactical call signs

E2C09 (C)
What type of equipment is commonly used to implement a ham radio mesh network?
A. A 2 meter VHF transceiver with a 1200 baud modem
B. An optical cable connection between the USB ports of 2 separate computers
C. A standard wireless router running custom software
D. A 440 MHz transceiver with a 9600 baud modem

E2C10 (D)
Why might a DX station state that they are listening on another frequency?
A. Because the DX station may be transmitting on a frequency that is prohibited to some responding stations
B. To separate the calling stations from the DX station
C. To improve operating efficiency by reducing interference
D. All of these choices are correct

E2C11 (A)
How should you generally identify your station when attempting to contact a DX station during a contest or in a pileup?
A. Send your full call sign once or twice
B. Send only the last two letters of your call sign until you make contact
C. Send your full call sign and grid square
D. Send the call sign of the DX station three times, the words "this is", then your call sign three times

E2C12 (B)
What might help to restore contact when DX signals become too weak to copy across an entire HF band a few hours after sunset?
A. Switch to a higher frequency HF band
B. Switch to a lower frequency HF band
C. Wait 90 minutes or so for the signal degradation to pass
D. Wait 24 hours before attempting another communication on the band

E2C13 (D)
What indicator is required to be used by U.S.-licensed operators when operating a station via remote control where the transmitter is located in the U.S.?
A. / followed by the USPS two letter abbreviation for the state in which the remote station is located
B. /R# where # is the district of the remote station
C. The ARRL section of the remote station
D. No additional indicator is required

E2D - OPERATING METHODS: VHF AND UHF DIGITAL MODES AND PROCEDURES; APRS; EME PROCEDURES, METEOR SCATTER PROCEDURES

E2D01 (B)
Which of the following digital modes is especially designed for use for meteor scatter signals?
A. WSPR
B. FSK441
C. Hellschreiber
D. APRS

E2D02 (D)
Which of the following is a good technique for making meteor scatter contacts?
A. 15 second timed transmission sequences with stations alternating based on location
B. Use of high speed CW or digital modes
C. Short transmission with rapidly repeated call signs and signal reports
D. All of these choices are correct

E2D03 (D)
Which of the following digital modes is especially useful for EME communications?
A. FSK441
B. PACTOR III
C. Olivia
D. JT65

E2D04 (C)
What is the purpose of digital store-and-forward functions on an Amateur Radio satellite?
A. To upload operational software for the transponder
B. To delay download of telemetry between satellites
C. To store digital messages in the satellite for later download by other stations
D. To relay messages between satellites

E2D05 (B)
Which of the following techniques is normally used by low Earth orbiting digital satellites to relay messages around the world?
A. Digipeating
B. Store-and-forward
C. Multi-satellite relaying
D. Node hopping

E2D06 (A)
Which of the following describes a method of establishing EME contacts?
A. Time synchronous transmissions alternately from each station
B. Storing and forwarding digital messages
C. Judging optimum transmission times by monitoring beacons reflected from the Moon
D. High speed CW identification to avoid fading

E2D07 (C)
What digital protocol is used by APRS?
A. PACTOR
B. 802.11
C. AX.25
D. AMTOR

E2D08 (A)
What type of packet frame is used to transmit APRS beacon data?
A. Unnumbered Information
B. Disconnect
C. Acknowledgement
D. Connect

E2D09 (D)
Which of these digital modes has the fastest data throughput under clear communication conditions?
A. AMTOR
B. 170 Hz shift, 45 baud RTTY
C. PSK31
D. 300 baud packet

E2D10 (C)
How can an APRS station be used to help support a public service communications activity?
A. An APRS station with an emergency medical technician can automatically transmit medical data to the nearest hospital
B. APRS stations with General Personnel Scanners can automatically relay the participant numbers and time as they pass the check points
C. An APRS station with a GPS unit can automatically transmit information to show a mobile station's position during the event
D. All of these choices are correct

E2D11 (D)
Which of the following data are used by the APRS network to communicate your location?
A. Polar coordinates
B. Time and frequency
C. Radio direction finding spectrum analysis
D. Latitude and longitude

E2D12 (A)
How does JT65 improve EME communications?
A. It can decode signals many dB below the noise floor using FEC
B. It controls the receiver to track Doppler shift
C. It supplies signals to guide the antenna to track the Moon
D. All of these choices are correct

E2D13 (A)
What type of modulation is used for JT65 contacts?
A. Multi-tone AFSK
B. PSK
C. RTTY
D. IEEE 802.11

E2D14 (B)
What is one advantage of using JT65 coding?
A. Uses only a 65 Hz bandwidth
B. The ability to decode signals which have a very low signal to noise ratio
C. Easily copied by ear if necessary
D. Permits fast-scan TV transmissions over narrow bandwidth

E2E - OPERATING METHODS: OPERATING HF DIGITAL MODES

E2E01 (B)
Which type of modulation is common for data emissions below 30 MHz?
A. DTMF tones modulating an FM signal
B. FSK
C. Pulse modulation
D. Spread spectrum

E2E02 (A)
What do the letters FEC mean as they relate to digital operation?
A. Forward Error Correction
B. First Error Correction
C. Fatal Error Correction
D. Final Error Correction

E2E03 (C)
How is the timing of JT65 contacts organized?
A. By exchanging ACK/NAK packets
B. Stations take turns on alternate days
C. Alternating transmissions at 1 minute intervals
D. It depends on the lunar phase

E2E04 (A)
What is indicated when one of the ellipses in an FSK crossed-ellipse display suddenly disappears?
A. Selective fading has occurred
B. One of the signal filters is saturated
C. The receiver has drifted 5 kHz from the desired receive frequency
D. The mark and space signal have been inverted

E2E05 (A)
Which type of digital mode does not support keyboard-to-keyboard operation?
A. Winlink
B. RTTY
C. PSK31
D. MFSK

E2E06 (C)
What is the most common data rate used for HF packet?
A. 48 baud
B. 110 baud
C. 300 baud
D. 1200 baud

E2E07 (B)
What is the typical bandwidth of a properly modulated MFSK16 signal?
A. 31 Hz
B. 316 Hz
C. 550 Hz
D. 2.16 kHz

E2E08 (B)
Which of the following HF digital modes can be used to transfer binary files?
A. Hellschreiber
B. PACTOR
C. RTTY
D. AMTOR

E2E09 (D)
Which of the following HF digital modes uses variable-length coding for bandwidth efficiency?
A. RTTY
B. PACTOR
C. MT63
D. PSK31

E2E10 (C)
Which of these digital modes has the narrowest bandwidth?
A. MFSK16
B. 170 Hz shift, 45 baud RTTY
C. PSK31
D. 300-baud packet

E2E11 (A)
What is the difference between direct FSK and audio FSK?
A. Direct FSK applies the data signal to the transmitter VFO
B. Audio FSK has a superior frequency response
C. Direct FSK uses a DC-coupled data connection
D. Audio FSK can be performed anywhere in the transmit chain

E2E12 (C)
Which type of control is used by stations using the Automatic Link Enable (ALE) protocol?
A. Local
B. Remote
C. Automatic
D. ALE can use any type of control

E2E13 (D)
Which of the following is a possible reason that attempts to initiate contact with a digital station on a clear frequency are unsuccessful?
A. Your transmit frequency is incorrect
B. The protocol version you are using is not the supported by the digital station
C. Another station you are unable to hear is using the frequency
D. All of these choices are correct

Subelement E3 - Radio Wave Propagation [3 Groups]

E3A - ELECTROMAGNETIC WAVES; EARTH-MOON-EARTH COMMUNICATIONS; METEOR
    SCATTER; MICROWAVE TROPOSPHERIC AND SCATTER PROPAGATION; AURORA
    PROPAGATION

E3A01 (D)
What is the approximate maximum separation measured along the surface of the Earth between two stations communicating by Moon bounce?
A. 500 miles, if the Moon is at perigee
B. 2000 miles, if the Moon is at apogee
C. 5000 miles, if the Moon is at perigee
D. 12,000 miles, if the Moon is visible by both stations

E3A02 (B)
What characterizes libration fading of an EME signal?
A. A slow change in the pitch of the CW signal
B. A fluttery irregular fading
C. A gradual loss of signal as the Sun rises
D. The returning echo is several Hertz lower in frequency than the transmitted signal

E3A03 (A)
When scheduling EME contacts, which of these conditions will generally result in the least path loss?
A. When the Moon is at perigee
B. When the Moon is full
C. When the Moon is at apogee
D. When the MUF is above 30 MHz

E3A04 (D)
What do Hepburn maps predict?
A. Sporadic E propagation
B. Locations of auroral reflecting zones
C. Likelihood of rain-scatter along cold or warm fronts
D. Probability of tropospheric propagation

E3A05 (C)
Tropospheric propagation of microwave signals often occurs along what weather related structure?
A. Gray-line
B. Lightning discharges
C. Warm and cold fronts
D. Sprites and jets

E3A06 (C)
Which of the following is required for microwave propagation via rain scatter?
A. Rain droplets must be electrically charged
B. Rain droplets must be within the E layer
C. The rain must be within radio range of both stations
D. All of these choices are correct

E3A07 (C)
Atmospheric ducts capable of propagating microwave signals often form over what geographic feature?
A. Mountain ranges
B. Forests
C. Bodies of water
D. Urban areas

E3A08 (A)
When a meteor strikes the Earth's atmosphere, a cylindrical region of free electrons is formed at what layer of the ionosphere?
A. The E layer
B. The F1 layer
C. The F2 layer
D. The D layer

E3A09 (C)
Which of the following frequency range is most suited for meteor scatter communications?
A. 1.8 MHz - 1.9 MHz
B. 10 MHz - 14 MHz
C. 28 MHz - 148 MHz
D. 220 MHz - 450 MHz

E3A10 (B)
Which type of atmospheric structure can create a path for microwave propagation?
A. The jet stream
B. Temperature inversion
C. Wind shear
D. Dust devil

E3A11 (B)
What is a typical range for tropospheric propagation of microwave signals?
A. 10 miles to 50 miles
B. 100 miles to 300 miles
C. 1200 miles
D. 2500 miles

E3A12 (C)
What is the cause of auroral activity?
A. The interaction in the F2 layer between the solar wind and the Van Allen belt
B. A low sunspot level combined with tropospheric ducting
C. The interaction in the E layer of charged particles from the Sun with the Earth's magnetic field
D. Meteor showers concentrated in the extreme northern and southern latitudes

E3A13 (A)
Which emission mode is best for aurora propagation?
A. CW
B. SSB
C. FM
D. RTTY

E3A14 (B)
From the contiguous 48 states, in which approximate direction should an antenna be pointed to take maximum advantage of aurora propagation?
A. South
B. North
C. East
D. West

E3A15 (C)
What is an electromagnetic wave?
A. A wave of alternating current, in the core of an electromagnet
B. A wave consisting of two electric fields at parallel right angles to each other
C. A wave consisting of an electric field and a magnetic field oscillating at right angles to each other
D. A wave consisting of two magnetic fields at right angles to each other

E3A16 (D)
Which of the following best describes electromagnetic waves traveling in free space?
A. Electric and magnetic fields become aligned as they travel
B. The energy propagates through a medium with a high refractive index
C. The waves are reflected by the ionosphere and return to their source
D. Changing electric and magnetic fields propagate the energy

E3A17 (B)
What is meant by circularly polarized electromagnetic waves?
A. Waves with an electric field bent into a circular shape
B. Waves with a rotating electric field
C. Waves that circle the Earth
D. Waves produced by a loop antenna

E3B - TRANSEQUATORIAL PROPAGATION; LONG PATH; GRAY-LINE; MULTI-PATH; ORDINARY AND EXTRAORDINARY WAVES; CHORDAL HOP, SPORADIC E MECHANISMS

E3B01 (A)
What is transequatorial propagation?
A. Propagation between two mid-latitude points at approximately the same distance north and south of the magnetic equator
B. Propagation between any two points located on the magnetic equator
C. Propagation between two continents by way of ducts along the magnetic equator
D. Propagation between two stations at the same latitude

E3B02 (C)
What is the approximate maximum range for signals using transequatorial propagation?
A. 1000 miles
B. 2500 miles
C. 5000 miles
D. 7500 miles

E3B03 (C)
What is the best time of day for transequatorial propagation?
A. Morning
B. Noon
C. Afternoon or early evening
D. Late at night

E3B04 (B)
What is meant by the terms extraordinary and ordinary waves?
A. Extraordinary waves describe rare long skip propagation compared to ordinary waves which travel shorter distances
B. Independent waves created in the ionosphere that are elliptically polarized
C. Long path and short path waves
D. Refracted rays and reflected waves

E3B05 (C)
Which amateur bands typically support long-path propagation?
A. 160 meters to 40 meters
B. 30 meters to 10 meters
C. 160 meters to 10 meters
D. 6 meters to 2 meters

E3B06 (B)
Which of the following amateur bands most frequently provides long-path propagation?
A. 80 meters
B. 20 meters
C. 10 meters
D. 6 meters

E3B07 (D)
Which of the following could account for hearing an echo on the received signal of a distant station?
A. High D layer absorption
B. Meteor scatter
C. Transmit frequency is higher than the MUF
D. Receipt of a signal by more than one path

E3B08 (D)
What type of HF propagation is probably occurring if radio signals travel along the terminator between daylight and darkness?
A. Transequatorial
B. Sporadic-E
C. Long-path
D. Gray-line

E3B09 (A)
At what time of year is Sporadic E propagation most likely to occur?
A. Around the solstices, especially the summer solstice
B. Around the solstices, especially the winter solstice
C. Around the equinoxes, especially the spring equinox
D. Around the equinoxes, especially the fall equinox

E3B10 (B)
What is the cause of gray-line propagation?
A. At midday, the Sun super heats the ionosphere causing increased refraction of radio waves
B. At twilight and sunrise, D-layer absorption is low while E-layer and F-layer propagation remains high
C. In darkness, solar absorption drops greatly while atmospheric ionization remains steady
D. At mid-afternoon, the Sun heats the ionosphere decreasing radio wave refraction and the MUF

E3B11 (D)
At what time of day is Sporadic-E propagation most likely to occur?
A. Around sunset
B. Around sunrise
C. Early evening
D. Any time

E3B12 (B)
What is the primary characteristic of chordal hop propagation?
A. Propagation away from the great circle bearing between stations
B. Successive ionospheric reflections without an intermediate reflection from the ground
C. Propagation across the geomagnetic equator
D. Signals reflected back toward the transmitting station

E3B13 (A)
Why is chordal hop propagation desirable?
A. The signal experiences less loss along the path compared to normal skip propagation
B. The MUF for chordal hop propagation is much lower than for normal skip propagation
C. Atmospheric noise is lower in the direction of chordal hop propagation
D. Signals travel faster along ionospheric chords

E3B14 (C)
What happens to linearly polarized radio waves that split into ordinary and extradinary waves in the ionosphere?
A. They are bent toward the magnetic poles
B. Their polarization is randomly modified
C. They become elliptically polarized
D. They become phase-locked

E3C - RADIO-PATH HORIZON; LESS COMMON PROPAGATION MODES; PROPAGATION PREDICTION TECHNIQUES AND MODELING; SPACE WEATHER PARAMETERS AND AMATEUR RADIO

E3C01 (B)
What does the term ray tracing describe in regard to radio communications?
A. The process in which an electronic display presents a pattern
B. Modeling a radio wave's path through the ionosphere
C. Determining the radiation pattern from an array of antennas
D. Evaluating high voltage sources for X-Rays

E3C02 (A)
What is indicated by a rising A or K index?
A. Increasing disruption of the geomagnetic field
B. Decreasing disruption of the geomagnetic field
C. Higher levels of solar UV radiation
D. An increase in the critical frequency

E3C03 (B)
Which of the following signal paths is most likely to experience high levels of absorption when the A index or K index is elevated?
A. Transequatorial propagation
B. Polar paths
C. Sporadic-E
D. NVIS

E3C04 (C)
What does the value of Bz (B sub Z) represent?
A. Geomagnetic field stability
B. Critical frequency for vertical transmissions
C. Direction and strength of the interplanetary magnetic field
D. Duration of long-delayed echoes

E3C05 (A)
What orientation of Bz (B sub z) increases the likelihood that incoming particles from the Sun will cause disturbed conditions?
A. Southward
B. Northward
C. Eastward
D. Westward

E3C06 (A)
By how much does the VHF/UHF radio horizon distance exceed the geometric horizon?
A. By approximately 15 percent of the distance
B. By approximately twice the distance
C. By approximately 50 percent of the distance
D. By approximately four times the distance

E3C07 (D)
Which of the following descriptors indicates the greatest solar flare intensity?
A. Class A
B. Class B
C. Class M
D. Class X

E3C08 (A)
What does the space weather term G5 mean?
A. An extreme geomagnetic storm
B. Very low solar activity
C. Moderate solar wind
D. Waning sunspot numbers

E3C09 (C)
How does the intensity of an X3 flare compare to that of an X2 flare?
A. 10 percent greater
B. 50 percent greater
C. Twice as great
D. Four times as great

E3C10 (B)
What does the 304A solar parameter measure?
A. The ratio of X-Ray flux to radio flux, correlated to sunspot number
B. UV emissions at 304 angstroms, correlated to solar flux index
C. The solar wind velocity at 304 degrees from the solar equator, correlated to solar activity
D. The solar emission at 304 GHz, correlated to X-Ray flare levels

E3C11 (C)
What does VOACAP software model?
A. AC voltage and impedance
B. VHF radio propagation
C. HF propagation
D. AC current and impedance

E3C12 (C)
How does the maximum distance of ground-wave propagation change when the signal frequency is increased?
A. It stays the same
B. It increases
C. It decreases
D. It peaks at roughly 14 MHz

E3C13 (A)
What type of polarization is best for ground-wave propagation?
A. Vertical
B. Horizontal
C. Circular
D. Elliptical

E3C14 (D)
Why does the radio-path horizon distance exceed the geometric horizon?
A. E-region skip
B. D-region skip
C. Downward bending due to aurora refraction
D. Downward bending due to density variations in the atmosphere

E3C15 (B)
What might a sudden rise in radio background noise indicate?
A. A meteor ping
B. A solar flare has occurred
C. Increased transequatorial propagation likely
D. Long-path propagation is occurring

Subelement E4 - Amateur Practices [5 Groups]

E4A - TEST EQUIPMENT: ANALOG AND DIGITAL INSTRUMENTS; SPECTRUM AND NETWORK ANALYZERS, ANTENNA ANALYZERS; OSCILLOSCOPES; RF MEASUREMENTS; COMPUTER AIDED MEASUREMENTS

E4A01 (C)
Which of the following parameter determines the bandwidth of a digital or computer-based oscilloscope?
A. Input capacitance
B. Input impedance
C. Sampling rate
D. Sample resolution

E4A02 (B)
Which of the following parameters would a spectrum analyzer display on the vertical and horizontal axes?
A. RF amplitude and time
B. RF amplitude and frequency
C. SWR and frequency
D. SWR and time

E4A03 (B)
Which of the following test instrument is used to display spurious signals and/or intermodulation distortion products in an SSB transmitter?
A. A wattmeter
B. A spectrum analyzer
C. A logic analyzer
D. A time-domain reflectometer

E4A04 (A)
What determines the upper frequency limit for a computer soundcard-based oscilloscope program?
A. Analog-to-digital conversion speed of the soundcard
B. Amount of memory on the soundcard
C. Q of the interface of the interface circuit
D. All of these choices are correct

E4A05 (D)
What might be an advantage of a digital vs an analog oscilloscope?
A. Automatic amplitude and frequency numerical readout
B. Storage of traces for future reference
C. Manipulation of time base after trace capture
D. All of these choices are correct

E4A06 (A)
What is the effect of aliasing in a digital or computer-based oscilloscope?
A. False signals are displayed
B. All signals will have a DC offset
C. Calibration of the vertical scale is no longer valid
D. False triggering occurs

E4A07 (B)
Which of the following is an advantage of using an antenna analyzer compared to an SWR bridge to measure antenna SWR?
A. Antenna analyzers automatically tune your antenna for resonance
B. Antenna analyzers do not need an external RF source
C. Antenna analyzers display a time-varying representation of the modulation envelope
D. All of these choices are correct

E4A08 (D)
Which of the following instrument would be best for measuring the SWR of a beam antenna?
A. A spectrum analyzer
B. A Q meter
C. An ohmmeter
D. An antenna analyzer

E4A09 (B)
When using a computer's soundcard input to digitize signals, what is the highest frequency signal that can be digitized without aliasing?
A. The same as the sample rate
B. One-half the sample rate
C. One-tenth the sample rate
D. It depends on how the data is stored internally

E4A10 (D)
Which of the following displays multiple digital signal states simultaneously?
A. Network analyzer
B. Bit error rate tester
C. Modulation monitor
D. Logic analyzer

E4A11 (A)
Which of the following is good practice when using an oscilloscope probe?
A. Keep the signal ground connection of the probe as short as possible
B. Never use a high impedance probe to measure a low impedance circuit
C. Never use a DC-coupled probe to measure an AC circuit
D. All of these choices are correct

E4A12 (B)
Which of the following procedures is an important precaution to follow when connecting a spectrum analyzer to a transmitter output?
A. Use high quality double shielded coaxial cables to reduce signal losses
B. Attenuate the transmitter output going to the spectrum analyzer
C. Match the antenna to the load
D. All of these choices are correct

E4A13 (A)
How is the compensation of an oscilloscope probe typically adjusted?
A. A square wave is displayed and the probe is adjusted until the horizontal portions of the displayed wave are as nearly flat as possible
B. A high frequency sine wave is displayed and the probe is adjusted for maximum amplitude
C. A frequency standard is displayed and the probe is adjusted until the deflection time is accurate
D. A DC voltage standard is displayed and the probe is adjusted until the displayed voltage is accurate

E4A14 (D)
What is the purpose of the prescaler function on a frequency counter?
A. It amplifies low level signals for more accurate counting
B. It multiplies a higher frequency signal so a low-frequency counter can display the operating frequency
C. It prevents oscillation in a low-frequency counter circuit
D. It divides a higher frequency signal so a low-frequency counter can display the input frequency

E4A15 (C)
What is an advantage of a period-measuring frequency counter over a direct-count type?
A. It can run on battery power for remote measurements
B. It does not require an expensive high-precision time base
C. It provides improved resolution of low-frequency signals within a comparable time period
D. It can directly measure the modulation index of an FM transmitter

E4B - MEASUREMENT TECHNIQUE AND LIMITATIONS: INSTRUMENT ACCURACY AND PERFORMANCE LIMITATIONS; PROBES; TECHNIQUES TO MINIMIZE ERRORS; MEASUREMENT OF "Q"; INSTRUMENT CALIBRATION; S PARAMETERS; VECTOR NETWORK ANALYZERS

E4B01 (B)
Which of the following factors most affects the accuracy of a frequency counter?
A. Input attenuator accuracy
B. Time base accuracy
C. Decade divider accuracy
D. Temperature coefficient of the logic

E4B02 (C)
What is an advantage of using a bridge circuit to measure impedance?
A. It provides an excellent match under all conditions
B. It is relatively immune to drift in the signal generator source
C. It is very precise in obtaining a signal null
D. It can display results directly in Smith chart format

E4B03 (C)
If a frequency counter with a specified accuracy of +/- 1.0 ppm reads 146,520,000 Hz, what is the most the actual frequency being measured could differ from the reading?
A. 165.2 Hz
B. 14.652 kHz
C. 146.52 Hz
D. 1.4652 MHz

E4B04 (A)
If a frequency counter with a specified accuracy of +/- 0.1 ppm reads 146,520,000 Hz, what is the most the actual frequency being measured could differ from the reading?
A. 14.652 Hz
B. 0.1 MHz
C. 1.4652 Hz
D. 1.4652 kHz

E4B05 (D)
If a frequency counter with a specified accuracy of +/- 10 ppm reads 146,520,000 Hz, what is the most the actual frequency being measured could differ from the reading?
A. 146.52 Hz
B. 10 Hz
C. 146.52 kHz
D. 1465.20 Hz

E4B06 (D)
How much power is being absorbed by the load when a directional power meter connected between a transmitter and a terminating load reads 100 watts forward power and 25 watts reflected power?
A. 100 watts
B. 125 watts
C. 25 watts
D. 75 watts

E4B07 (A)
What do the subscripts of S parameters represent?
A. The port or ports at which measurements are made
B. The relative time between measurements
C. Relative quality of the data
D. Frequency order of the measurements

E4B08 (C)
Which of the following is a characteristic of a good DC voltmeter?
A. High reluctance input
B. Low reluctance input
C. High impedance input
D. Low impedance input

E4B09 (D)
What is indicated if the current reading on an RF ammeter placed in series with the antenna feed line of a transmitter increases as the transmitter is tuned to resonance?
A. There is possibly a short to ground in the feed line
B. The transmitter is not properly neutralized
C. There is an impedance mismatch between the antenna and feed line
D. There is more power going into the antenna

E4B10 (B)
Which of the following describes a method to measure intermodulation distortion in an SSB transmitter?
A. Modulate the transmitter with two non-harmonically related radio frequencies and observe the RF output with a spectrum analyzer
B. Modulate the transmitter with two non-harmonically related audio frequencies and observe the RF output with a spectrum analyzer
C. Modulate the transmitter with two harmonically related audio frequencies and observe the RF output with a peak reading wattmeter
D. Modulate the transmitter with two harmonically related audio frequencies and observe the RF output with a logic analyzer

E4B11 (D)
How should an antenna analyzer be connected when measuring antenna resonance and feed point impedance?
A. Loosely couple the analyzer near the antenna base
B. Connect the analyzer via a high-impedance transformer to the antenna
C. Loosely couple the antenna and a dummy load to the analyzer
D. Connect the antenna feed line directly to the analyzer's connector

E4B12 (A)
What is the significance of voltmeter sensitivity expressed in ohms per volt?
A. The full scale reading of the voltmeter multiplied by its ohms per volt rating will indicate the input impedance of the voltmeter
B. When used as a galvanometer, the reading in volts multiplied by the ohms per volt rating will determine the power drawn by the device under test
C. When used as an ohmmeter, the reading in ohms divided by the ohms per volt rating will determine the voltage applied to the circuit
D. When used as an ammeter, the full scale reading in amps divided by ohms per volt rating will determine the size of shunt needed

E4B13 (C)
Which S parameter is equivalent to forward gain?
A. S11
B. S12
C. S21
D. S22

E4B14 (B)
What happens if a dip meter is too tightly coupled to a tuned circuit being checked?
A. Harmonics are generated
B. A less accurate reading results
C. Cross modulation occurs
D. Intermodulation distortion occurs

E4B15 (C)
Which of the following can be used as a relative measurement of the Q for a series-tuned circuit?
A. The inductance to capacitance ratio
B. The frequency shift
C. The bandwidth of the circuit's frequency response
D. The resonant frequency of the circuit

E4B16 (A)
Which S parameter represents return loss or SWR?
A. S11
B. S12
C. S21
D. S22

E4B17 (B)
What three test loads are used to calibrate a standard RF vector network analyzer?
A. 50 ohms, 75 ohms, and 90 ohms
B. Short circuit, open circuit, and 50 ohms
C. Short circuit, open circuit, and resonant circuit
D. 50 ohms through 1/8 wavelength, 1/4 wavelength, and 1/2 wavelength of coaxial cable

E4C - RECEIVER PERFORMANCE CHARACTERISTICS, PHASE NOISE, NOISE FLOOR, IMAGE REJECTION, MDS, SIGNAL-TO-NOISE-RATIO; SELECTIVITY; EFFECTS OF SDR RECEIVER NON-LINEARITY

E4C01 (D)
What is an effect of excessive phase noise in the local oscillator section of a receiver?
A. It limits the receiver's ability to receive strong signals
B. It reduces receiver sensitivity
C. It decreases receiver third-order intermodulation distortion dynamic range
D. It can cause strong signals on nearby frequencies to interfere with reception of weak signals

E4C02 (A)
Which of the following portions of a receiver can be effective in eliminating image signal interference?
A. A front-end filter or pre-selector
B. A narrow IF filter
C. A notch filter
D. A properly adjusted product detector

E4C03 (C)
What is the term for the blocking of one FM phone signal by another, stronger FM phone signal?
A. Desensitization
B. Cross-modulation interference
C. Capture effect
D. Frequency discrimination

E4C04 (D)
How is the noise figure of a receiver defined?
A. The ratio of atmospheric noise to phase noise
B. The ratio of the noise bandwidth in Hertz to the theoretical bandwidth of a resistive network
C. The ratio of thermal noise to atmospheric noise
D. The ratio in dB of the noise generated by the receiver to the theoretical minimum noise

E4C05 (B)
What does a value of -174 dBm/Hz represent with regard to the noise floor of a receiver?
A. The minimum detectable signal as a function of receive frequency
B. The theoretical noise at the input of a perfect receiver at room temperature
C. The noise figure of a 1 Hz bandwidth receiver
D. The galactic noise contribution to minimum detectable signal

E4C06 (D)
A CW receiver with the AGC off has an equivalent input noise power density of -174 dBm/Hz. What would be the level of an unmodulated carrier input to this receiver that would yield an audio output SNR of 0 dB in a 400 Hz noise bandwidth?
A. -174 dBm
B. -164 dBm
C. -155 dBm
D. -148 dBm

E4C07 (B)
What does the MDS of a receiver represent?
A. The meter display sensitivity
B. The minimum discernible signal
C. The multiplex distortion stability
D. The maximum detectable spectrum

E4C08 (C)
An SDR receiver is overloaded when input signals exceed what level?
A. One-half the maximum sample rate
B. One-half the maximum sampling buffer size
C. The maximum count value of the analog-to-digital converter
D. The reference voltage of the analog-to-digital converter

E4C09 (C)
Which of the following choices is a good reason for selecting a high frequency for the design of the IF in a conventional HF or VHF communications receiver?
A. Fewer components in the receiver
B. Reduced drift
C. Easier for front-end circuitry to eliminate image responses
D. Improved receiver noise figure

E4C10 (B)
Which of the following is a desirable amount of selectivity for an amateur RTTY HF receiver?
A. 100 Hz
B. 300 Hz
C. 6000 Hz
D. 2400 Hz

E4C11 (B)
Which of the following is a desirable amount of selectivity for an amateur SSB phone receiver?
A. 1 kHz
B. 2.4 kHz
C. 4.2 kHz
D. 4.8 kHz

E4C12 (D)
What is an undesirable effect of using too wide a filter bandwidth in the IF section of a receiver?
A. Output-offset overshoot
B. Filter ringing
C. Thermal-noise distortion
D. Undesired signals may be heard

E4C13 (C)
How does a narrow-band roofing filter affect receiver performance?
A. It improves sensitivity by reducing front end noise
B. It improves intelligibility by using low Q circuitry to reduce ringing
C. It improves dynamic range by attenuating strong signals near the receive frequency
D. All of these choices are correct

E4C14 (D)
What transmit frequency might generate an image response signal in a receiver tuned to 14.300 MHz and which uses a 455 kHz IF frequency?
A. 13.845 MHz
B. 14.755 MHz
C. 14.445 MHz
D. 15.210 MHz

E4C15 (D)
What is usually the primary source of noise that is heard from an HF receiver with an antenna connected?
A. Detector noise
B. Induction motor noise
C. Receiver front-end noise
D. Atmospheric noise

E4C16 (A)
Which of the following is caused by missing codes in an SDR receiver's analog-to-digital converter?
A. Distortion
B. Overload
C. Loss of sensitivity
D. Excess output level

E4C17 (D)
Which of the following has the largest effect on an SDR receiver's linearity?
A. CPU register width in bits
B. Anti-aliasing input filter bandwidth
C. RAM speed used for data storage
D. Analog-to-digital converter sample width in bits

E4D - RECEIVER PERFORMANCE CHARACTERISTICS: BLOCKING DYNAMIC RANGE; INTERMODULATION AND CROSS-MODULATION INTERFERENCE; 3RD ORDER INTERCEPT; DESENSITIZATION; PRESELECTOR

E4D01 (A)
What is meant by the blocking dynamic range of a receiver?
A. The difference in dB between the noise floor and the level of an incoming signal which will cause 1 dB of gain compression
B. The minimum difference in dB between the levels of two FM signals which will cause one signal to block the other
C. The difference in dB between the noise floor and the third order intercept point
D. The minimum difference in dB between two signals which produce third order intermodulation products greater than the noise floor

E4D02 (A)
Which of the following describes two problems caused by poor dynamic range in a communications receiver?
A. Cross-modulation of the desired signal and desensitization from strong adjacent signals
B. Oscillator instability requiring frequent retuning and loss of ability to recover the opposite sideband
C. Cross-modulation of the desired signal and insufficient audio power to operate the speaker
D. Oscillator instability and severe audio distortion of all but the strongest received signals

E4D03 (B)
How can intermodulation interference between two repeaters occur?
A. When the repeaters are in close proximity and the signals cause feedback in the final amplifier of one or both transmitters
B. When the repeaters are in close proximity and the signals mix in the final amplifier of one or both transmitters
C. When the signals from the transmitters are reflected out of phase from airplanes passing overhead
D. When the signals from the transmitters are reflected in phase from airplanes passing overhead

E4D04 (B)
Which of the following may reduce or eliminate intermodulation interference in a repeater caused by another transmitter operating in close proximity?
A. A band-pass filter in the feed line between the transmitter and receiver
B. A properly terminated circulator at the output of the transmitter
C. A Class C final amplifier
D. A Class D final amplifier

E4D05 (A)
What transmitter frequencies would cause an intermodulation-product signal in a receiver tuned to 146.70 MHz when a nearby station transmits on 146.52 MHz?
A. 146.34 MHz and 146.61 MHz
B. 146.88 MHz and 146.34 MHz
C. 146.10 MHz and 147.30 MHz
D. 173.35 MHz and 139.40 MHz

E4D06 (D)
What is the term for unwanted signals generated by the mixing of two or more signals?
A. Amplifier desensitization
B. Neutralization
C. Adjacent channel interference
D. Intermodulation interference

E4D07 (D)
Which describes the most significant effect of an off-frequency signal when it is causing cross-modulation interference to a desired signal?
A. A large increase in background noise
B. A reduction in apparent signal strength
C. The desired signal can no longer be heard
D. The off-frequency unwanted signal is heard in addition to the desired signal

E4D08 (C)
What causes intermodulation in an electronic circuit?
A. Too little gain
B. Lack of neutralization
C. Nonlinear circuits or devices
D. Positive feedback

E4D09 (C)
What is the purpose of the preselector in a communications receiver?
A. To store often-used frequencies
B. To provide a range of AGC time constants
C. To increase rejection of unwanted signals
D. To allow selection of the optimum RF amplifier device

E4D10 (C)
What does a third-order intercept level of 40 dBm mean with respect to receiver performance?
A. Signals less than 40 dBm will not generate audible third-order intermodulation products
B. The receiver can tolerate signals up to 40 dB above the noise floor without producing third-order intermodulation products
C. A pair of 40 dBm signals will theoretically generate a third-order intermodulation product with the same level as the input signals
D. A pair of 1 mW input signals will produce a third-order intermodulation product which is 40 dB stronger than the input signal

E4D11 (A)
Why are third-order intermodulation products created within a receiver of particular interest compared to other products?
A. The third-order product of two signals which are in the band of interest is also likely to be within the band
B. The third-order intercept is much higher than other orders
C. Third-order products are an indication of poor image rejection
D. Third-order intermodulation produces three products for every input signal within the band of interest

E4D12 (A)
What is the term for the reduction in receiver sensitivity caused by a strong signal near the received frequency?
A. Desensitization
B. Quieting
C. Cross-modulation interference
D. Squelch gain rollback

E4D13 (B)
Which of the following can cause receiver desensitization?
A. Audio gain adjusted too low
B. Strong adjacent channel signals
C. Audio bias adjusted too high
D. Squelch gain misadjusted

E4D14 (A)
Which of the following is a way to reduce the likelihood of receiver desensitization?
A. Decrease the RF bandwidth of the receiver
B. Raise the receiver IF frequency
C. Increase the receiver front end gain
D. Switch from fast AGC to slow AGC

E4E - NOISE SUPPRESSION: SYSTEM NOISE; ELECTRICAL APPLIANCE NOISE; LINE NOISE; LOCATING NOISE SOURCES; DSP NOISE REDUCTION; NOISE BLANKERS; GROUNDING FOR SIGNALS

E4E01 (A)
Which of the following types of receiver noise can often be reduced by use of a receiver noise blanker?
A. Ignition noise
B. Broadband white noise
C. Heterodyne interference
D. All of these choices are correct

E4E02 (D)
Which of the following types of receiver noise can often be reduced with a DSP noise filter?
A. Broadband white noise
B. Ignition noise
C. Power line noise
D. All of these choices are correct

E4E03 (B)
Which of the following signals might a receiver noise blanker be able to remove from desired signals?
A. Signals which are constant at all IF levels
B. Signals which appear across a wide bandwidth
C. Signals which appear at one IF but not another
D. Signals which have a sharply peaked frequency distribution

E4E04 (D)
How can conducted and radiated noise caused by an automobile alternator be suppressed?
A. By installing filter capacitors in series with the DC power lead and a blocking capacitor in the field lead
B. By installing a noise suppression resistor and a blocking capacitor in both leads
C. By installing a high-pass filter in series with the radio's power lead and a low-pass filter in parallel with the field lead
D. By connecting the radio's power leads directly to the battery and by installing coaxial capacitors in line with the alternator leads

E4E05 (B)
How can noise from an electric motor be suppressed?
A. By installing a high pass filter in series with the motor's power leads
B. By installing a brute-force AC-line filter in series with the motor leads
C. By installing a bypass capacitor in series with the motor leads
D. By using a ground-fault current interrupter in the circuit used to power the motor

E4E06 (B)
What is a major cause of atmospheric static?
A. Solar radio frequency emissions
B. Thunderstorms
C. Geomagnetic storms
D. Meteor showers

E4E07 (C)
How can you determine if line noise interference is being generated within your home?
A. By checking the power line voltage with a time domain reflectometer
B. By observing the AC power line waveform with an oscilloscope
C. By turning off the AC power line main circuit breaker and listening on a battery operated radio
D. By observing the AC power line voltage with a spectrum analyzer

E4E08 (A)
What type of signal is picked up by electrical wiring near a radio antenna?
A. A common-mode signal at the frequency of the radio transmitter
B. An electrical-sparking signal
C. A differential-mode signal at the AC power line frequency
D. Harmonics of the AC power line frequency

E4E09 (C)
What undesirable effect can occur when using an IF noise blanker?
A. Received audio in the speech range might have an echo effect
B. The audio frequency bandwidth of the received signal might be compressed
C. Nearby signals may appear to be excessively wide even if they meet emission standards
D. FM signals can no longer be demodulated

E4E10 (D)
What is a common characteristic of interference caused by a touch controlled electrical device?
A. The interfering signal sounds like AC hum on an AM receiver or a carrier modulated by 60 Hz hum on a SSB or CW receiver
B. The interfering signal may drift slowly across the HF spectrum
C. The interfering signal can be several kHz in width and usually repeats at regular intervals across a HF band
D. All of these choices are correct

E4E11 (B)
Which is the most likely cause if you are hearing combinations of local AM broadcast signals within one or more of the MF or HF ham bands?
A. The broadcast station is transmitting an over-modulated signal
B. Nearby corroded metal joints are mixing and re-radiating the broadcast signals
C. You are receiving sky wave signals from a distant station
D. Your station receiver IF amplifier stage is defective

E4E12 (A)
What is one disadvantage of using some types of automatic DSP notch-filters when attempting to copy CW signals?
A. A DSP filter can remove the desired signal at the same time as it removes interfering signals
B. Any nearby signal passing through the DSP system will overwhelm the desired signal
C. Received CW signals will appear to be modulated at the DSP clock frequency
D. Ringing in the DSP filter will completely remove the spaces between the CW characters

E4E13 (D)
What might be the cause of a loud roaring or buzzing AC line interference that comes and goes at intervals?
A. Arcing contacts in a thermostatically controlled device
B. A defective doorbell or doorbell transformer inside a nearby residence
C. A malfunctioning illuminated advertising display
D. All of these choices are correct

E4E14 (C)
What is one type of electrical interference that might be caused by the operation of a nearby personal computer?
A. A loud AC hum in the audio output of your station receiver
B. A clicking noise at intervals of a few seconds
C. The appearance of unstable modulated or unmodulated signals at specific frequencies
D. A whining type noise that continually pulses off and on

E4E15 (B)
Which of the following can cause shielded cables to radiate or receive interference?
A. Low inductance ground connections at both ends of the shield
B. Common mode currents on the shield and conductors
C. Use of braided shielding material
D. Tying all ground connections to a common point resulting in differential mode currents in the shield

E4E16 (B)
What current flows equally on all conductors of an unshielded multi-conductor cable?
A. Differential-mode current
B. Common-mode current
C. Reactive current only
D. Return current

Subelement E5 - Electrical Principles [4 Groups]

E5A - RESONANCE AND Q: CHARACTERISTICS OF RESONANT CIRCUITS: SERIES AND PARALLEL RESONANCE; DEFINITIONS AND EFFECTS OF Q; HALF-POWER BANDWIDTH; PHASE RELATIONSHIPS IN REACTIVE CIRCUITS

E5A01 (A)
What can cause the voltage across reactances in series to be larger than the voltage applied to them?
A. Resonance
B. Capacitance
C. Conductance
D. Resistance

E5A02 (C)
What is resonance in an electrical circuit?
A. The highest frequency that will pass current
B. The lowest frequency that will pass current
C. The frequency at which the capacitive reactance equals the inductive reactance
D. The frequency at which the reactive impedance equals the resistive impedance

E5A03 (D)
What is the magnitude of the impedance of a series RLC circuit at resonance?
A. High, as compared to the circuit resistance
B. Approximately equal to capacitive reactance
C. Approximately equal to inductive reactance
D. Approximately equal to circuit resistance

E5A04 (A)
What is the magnitude of the impedance of a circuit with a resistor, an inductor and a capacitor all in parallel, at resonance?
A. Approximately equal to circuit resistance
B. Approximately equal to inductive reactance
C. Low, as compared to the circuit resistance
D. Approximately equal to capacitive reactance

E5A05 (B)
What is the magnitude of the current at the input of a series RLC circuit as the frequency goes through resonance?
A. Minimum
B. Maximum
C. R/L
D. L/R

E5A06 (B)
What is the magnitude of the circulating current within the components of a parallel LC circuit at resonance?
A. It is at a minimum
B. It is at a maximum
C. It equals 1 divided by the quantity 2 times Pi, multiplied by the square root of inductance L multiplied by capacitance C
D. It equals 2 multiplied by Pi, multiplied by frequency, multiplied by inductance

E5A07 (A)
What is the magnitude of the current at the input of a parallel RLC circuit at resonance?
A. Minimum
B. Maximum
C. R/L
D. L/R

E5A08 (C)
What is the phase relationship between the current through and the voltage across a series resonant circuit at resonance?
A. The voltage leads the current by 90 degrees
B. The current leads the voltage by 90 degrees
C. The voltage and current are in phase
D. The voltage and current are 180 degrees out of phase

E5A09 (C)
How is the Q of an RLC parallel resonant circuit calculated?
A. Reactance of either the inductance or capacitance divided by the resistance
B. Reactance of either the inductance or capacitance multiplied by the resistance
C. Resistance divided by the reactance of either the inductance or capacitance
D. Reactance of the inductance multiplied by the reactance of the capacitance

E5A10 (A)
How is the Q of an RLC series resonant circuit calculated?
A. Reactance of either the inductance or capacitance divided by the resistance
B. Reactance of either the inductance or capacitance times the resistance
C. Resistance divided by the reactance of either the inductance or capacitance
D. Reactance of the inductance times the reactance of the capacitance

E5A11 (C)
What is the half-power bandwidth of a parallel resonant circuit that has a resonant frequency of 7.1 MHz and a Q of 150?
A. 157.8 Hz
B. 315.6 Hz
C. 47.3 kHz
D. 23.67 kHz

E5A12 (C)
What is the half-power bandwidth of a parallel resonant circuit that has a resonant frequency of 3.7 MHz and a Q of 118?
A. 436.6 kHz
B. 218.3 kHz
C. 31.4 kHz
D. 15.7 kHz

E5A13 (C)
What is an effect of increasing Q in a resonant circuit?
A. Fewer components are needed for the same performance
B. Parasitic effects are minimized
C. Internal voltages and circulating currents increase
D. Phase shift can become uncontrolled

E5A14 (C)
What is the resonant frequency of a series RLC circuit if R is 22 ohms, L is 50 microhenrys and C is 40 picofarads?
A. 44.72 MHz
B. 22.36 MHz
C. 3.56 MHz
D. 1.78 MHz

E5A15 (A)
Which of the following can increase Q for inductors and capacitors?
A. Lower losses
B. Lower reactance
C. Lower self-resonant frequency
D. Higher self-resonant frequency

E5A16 (D)
What is the resonant frequency of a parallel RLC circuit if R is 33 ohms, L is 50 microhenrys and C is 10 picofarads?
A. 23.5 MHz
B. 23.5 kHz
C. 7.12 kHz
D. 7.12 MHz

E5A17 (A)
What is the result of increasing the Q of an impedance-matching circuit?
A. Matching bandwidth is decreased
B. Matching bandwidth is increased
C. Matching range is increased
D. It has no effect on impedance matching

E5B - TIME CONSTANTS AND PHASE RELATIONSHIPS: RLC TIME CONSTANTS;
    DEFINITION; TIME CONSTANTS IN RL AND RC CIRCUITS; PHASE ANGLE BETWEEN
    VOLTAGE AND CURRENT; PHASE ANGLES OF SERIES RLC; PHASE ANGLE OF
    INDUCTANCE VS SUSCEPTANCE; ADMITTANCE AND SUSCEPTANCE

E5B01 (B)
What is the term for the time required for the capacitor in an RC circuit to be
charged to 63.2% of the applied voltage?
A. An exponential rate of one
B. One time constant
C. One exponential period
D. A time factor of one

E5B02 (D)
What is the term for the time it takes for a charged capacitor in an RC circuit to
discharge to 36.8% of its initial voltage?
A. One discharge period
B. An exponential discharge rate of one
C. A discharge factor of one
D. One time constant

E5B03 (B)
What happens to the phase angle of a reactance when it is converted to a
susceptance?
A. It is unchanged
B. The sign is reversed
C. It is shifted by 90 degrees
D. The susceptance phase angle is the inverse of the reactance phase angle

E5B04 (D)
What is the time constant of a circuit having two 220 microfarad capacitors and two
1 megohm resistors, all in parallel?
A. 55 seconds
B. 110 seconds
C. 440 seconds
D. 220 seconds

E5B05 (D)
What happens to the magnitude of a reactance when it is converted to a susceptance?
A. It is unchanged
B. The sign is reversed
C. It is shifted by 90 degrees
D. The magnitude of the susceptance is the reciprocal of the magnitude of the reactance

E5B06 (C)
What is susceptance?
A. The magnetic impedance of a circuit
B. The ratio of magnetic field to electric field
C. The inverse of reactance
D. A measure of the efficiency of a transformer

E5B07 (C)
What is the phase angle between the voltage across and the current through a series RLC circuit if XC is 500 ohms, R is 1 kilohm, and XL is 250 ohms?
A. 68.2 degrees with the voltage leading the current
B. 14.0 degrees with the voltage leading the current
C. 14.0 degrees with the voltage lagging the current
D. 68.2 degrees with the voltage lagging the current

E5B08 (A)
What is the phase angle between the voltage across and the current through a series RLC circuit if XC is 100 ohms, R is 100 ohms, and XL is 75 ohms?
A. 14 degrees with the voltage lagging the current
B. 14 degrees with the voltage leading the current
C. 76 degrees with the voltage leading the current
D. 76 degrees with the voltage lagging the current

E5B09 (D)
What is the relationship between the current through a capacitor and the voltage across a capacitor?
A. Voltage and current are in phase
B. Voltage and current are 180 degrees out of phase
C. Voltage leads current by 90 degrees
D. Current leads voltage by 90 degrees

E5B10 (A)
What is the relationship between the current through an inductor and the voltage across an inductor?
A. Voltage leads current by 90 degrees
B. Current leads voltage by 90 degrees
C. Voltage and current are 180 degrees out of phase
D. Voltage and current are in phase

E5B11 (B)
What is the phase angle between the voltage across and the current through a series RLC circuit if XC is 25 ohms, R is 100 ohms, and XL is 50 ohms?
A. 14 degrees with the voltage lagging the current
B. 14 degrees with the voltage leading the current
C. 76 degrees with the voltage lagging the current
D. 76 degrees with the voltage leading the current

E5B12 (A)
What is admittance?
A. The inverse of impedance
B. The term for the gain of a field effect transistor
C. The turns ratio of a transformer
D. The unit used for Q factor

E5B13 (D)
What letter is commonly used to represent susceptance?
A. G
B. X
C. Y
D. B

E5C - COORDINATE SYSTEMS AND PHASORS IN ELECTRONICS: RECTANGULAR COORDINATES; POLAR COORDINATES; PHASORS

E5C01 (A)
Which of the following represents a capacitive reactance in rectangular notation?
A. –jX
B. +jX
C. X
D. Omega

E5C02 (C)
How are impedances described in polar coordinates?
A. By X and R values
B. By real and imaginary parts
C. By phase angle and amplitude
D. By Y and G values

E5C03 (C)
Which of the following represents an inductive reactance in polar coordinates?
A. A positive real part
B. A negative real part
C. A positive phase angle
D. A negative phase angle

E5C04 (D)
Which of the following represents a capacitive reactance in polar coordinates?
A. A positive real part
B. A negative real part
C. A positive phase angle
D. A negative phase angle

E5C05 (C)
What is the name of the diagram used to show the phase relationship between impedances at a given frequency?
A. Venn diagram
B. Near field diagram
C. Phasor diagram
D. Far field diagram

E5C06 (B)
What does the impedance 50–j25 represent?
A. 50 ohms resistance in series with 25 ohms inductive reactance
B. 50 ohms resistance in series with 25 ohms capacitive reactance
C. 25 ohms resistance in series with 50 ohms inductive reactance
D. 25 ohms resistance in series with 50 ohms capacitive reactance

E5C07 (B)
What is a vector?
A. The value of a quantity that changes over time
B. A quantity with both magnitude and an angular component
C. The inverse of the tangent function
D. The inverse of the sine function

E5C08 (D)
What coordinate system is often used to display the phase angle of a circuit containing resistance, inductive and/or capacitive reactance?
A. Maidenhead grid
B. Faraday grid
C. Elliptical coordinates
D. Polar coordinates

E5C09 (A)
When using rectangular coordinates to graph the impedance of a circuit, what does the horizontal axis represent?
A. Resistive component
B. Reactive component
C. The sum of the reactive and resistive components
D. The difference between the resistive and reactive components

E5C10 (B)
When using rectangular coordinates to graph the impedance of a circuit, what does the vertical axis represent?
A. Resistive component
B. Reactive component
C. The sum of the reactive and resistive components
D. The difference between the resistive and reactive components

E5C11 (C)
What do the two numbers that are used to define a point on a graph using rectangular coordinates represent?
A. The magnitude and phase of the point
B. The sine and cosine values
C. The coordinate values along the horizontal and vertical axes
D. The tangent and cotangent values

E5C12 (D)
If you plot the impedance of a circuit using the rectangular coordinate system and find the impedance point falls on the right side of the graph on the horizontal axis, what do you know about the circuit?
A. It has to be a direct current circuit
B. It contains resistance and capacitive reactance
C. It contains resistance and inductive reactance
D. It is equivalent to a pure resistance

E5C13 (D)
What coordinate system is often used to display the resistive, inductive, and/or capacitive reactance components of impedance?
A. Maidenhead grid
B. Faraday grid
C. Elliptical coordinates
D. Rectangular coordinates

E5C14 (B)
Which point on Figure E5-2 best represents the impedance of a series circuit consisting of a 400 ohm resistor and a 38 picofarad capacitor at 14 MHz?
A. Point 2
B. Point 4
C. Point 5
D. Point 6

E5C15 (B)
Which point in Figure E5-2 best represents the impedance of a series circuit consisting of a 300 ohm resistor and an 18 microhenry inductor at 3.505 MHz?
A. Point 1
B. Point 3
C. Point 7
D. Point 8

E5C16 (A)
Which point on Figure E5-2 best represents the impedance of a series circuit consisting of a 300 ohm resistor and a 19 picofarad capacitor at 21.200 MHz?
A. Point 1
B. Point 3
C. Point 7
D. Point 8

E5C17 (D)
Which point on Figure E5-2 best represents the impedance of a series circuit consisting of a 300 ohm resistor, a 0.64-microhenry inductor and an 85-picofarad capacitor at 24.900 MHz?
A. Point 1
B. Point 3
C. Point 5
D. Point 8

E5D - AC AND RF ENERGY IN REAL CIRCUITS: SKIN EFFECT; ELECTROSTATIC AND ELECTROMAGNETIC FIELDS; REACTIVE POWER; POWER FACTOR; ELECTRICAL LENGTH OF CONDUCTORS AT UHF AND MICROWAVE FREQUENCIES

E5D01 (A)
What is the result of skin effect?
A. As frequency increases, RF current flows in a thinner layer of the conductor, closer to the surface
B. As frequency decreases, RF current flows in a thinner layer of the conductor, closer to the surface
C. Thermal effects on the surface of the conductor increase the impedance
D. Thermal effects on the surface of the conductor decrease the impedance

E5D02 (B)
Why is it important to keep lead lengths short for components used in circuits for VHF and above?
A. To increase the thermal time constant
B. To avoid unwanted inductive reactance
C. To maintain component lifetime
D. All of these choices are correct

E5D03 (D)
What is microstrip?
A. Lightweight transmission line made of common zip cord
B. Miniature coax used for low power applications
C. Short lengths of coax mounted on printed circuit boards to minimize time delay between microwave circuits
D. Precision printed circuit conductors above a ground plane that provide constant impedance interconnects at microwave frequencies

E5D04 (B)
Why are short connections necessary at microwave frequencies?
A. To increase neutralizing resistance
B. To reduce phase shift along the connection
C. Because of ground reflections
D. To reduce noise figure

E5D05 (A)
Which parasitic characteristic increases with conductor length?
A. Inductance
B. Permeability
C. Permittivity
D. Malleability

E5D06 (D)
In what direction is the magnetic field oriented about a conductor in relation to the direction of electron flow?
A. In the same direction as the current
B. In a direction opposite to the current
C. In all directions; omni-directional
D. In a direction determined by the left-hand rule

E5D07 (D)
What determines the strength of the magnetic field around a conductor?
A. The resistance divided by the current
B. The ratio of the current to the resistance
C. The diameter of the conductor
D. The amount of current flowing through the conductor

E5D08 (B)
What type of energy is stored in an electromagnetic or electrostatic field?
A. Electromechanical energy
B. Potential energy
C. Thermodynamic energy
D. Kinetic energy

E5D09 (B)
What happens to reactive power in an AC circuit that has both ideal inductors and ideal capacitors?
A. It is dissipated as heat in the circuit
B. It is repeatedly exchanged between the associated magnetic and electric fields, but is not dissipated
C. It is dissipated as kinetic energy in the circuit
D. It is dissipated in the formation of inductive and capacitive fields

E5D10 (A)
How can the true power be determined in an AC circuit where the voltage and current are out of phase?
A. By multiplying the apparent power times the power factor
B. By dividing the reactive power by the power factor
C. By dividing the apparent power by the power factor
D. By multiplying the reactive power times the power factor

E5D11 (C)
What is the power factor of an R-L circuit having a 60 degree phase angle between the voltage and the current?
A. 1.414
B. 0.866
C. 0.5
D. 1.73

E5D12 (B)
How many watts are consumed in a circuit having a power factor of 0.2 if the input is 100-VAC at 4 amperes?
A. 400 watts
B. 80 watts
C. 2000 watts
D. 50 watts

E5D13 (B)
How much power is consumed in a circuit consisting of a 100 ohm resistor in series with a 100 ohm inductive reactance drawing 1 ampere?
A. 70.7 Watts
B. 100 Watts
C. 141.4 Watts
D. 200 Watts

E5D14 (A)
What is reactive power?
A. Wattless, nonproductive power
B. Power consumed in wire resistance in an inductor
C. Power lost because of capacitor leakage
D. Power consumed in circuit Q

E5D15 (D)
What is the power factor of an R-L circuit having a 45 degree phase angle between the voltage and the current?
A. 0.866
B. 1.0
C. 0.5
D. 0.707

E5D16 (C)
What is the power factor of an R-L circuit having a 30 degree phase angle between the voltage and the current?
A. 1.73
B. 0.5
C. 0.866
D. 0.577

E5D17 (D)
How many watts are consumed in a circuit having a power factor of 0.6 if the input is 200VAC at 5 amperes?
A. 200 watts
B. 1000 watts
C. 1600 watts
D. 600 watts

E5D18 (B)
How many watts are consumed in a circuit having a power factor of 0.71 if the apparent power is 500VA?
A. 704 W
B. 355 W
C. 252 W
D. 1.42 mW

Subelement E6 - Circuit Components [6 Groups]

E6A - SEMICONDUCTOR MATERIALS AND DEVICES: SEMICONDUCTOR MATERIALS; GERMANIUM, SILICON, P-TYPE, N-TYPE; TRANSISTOR TYPES: NPN, PNP, JUNCTION, FIELD-EFFECT TRANSISTORS: ENHANCEMENT MODE; DEPLETION MODE; MOS; CMOS; N-CHANNEL; P-CHANNEL

E6A01 (C)
In what application is gallium arsenide used as a semiconductor material in preference to germanium or silicon?
A. In high-current rectifier circuits
B. In high-power audio circuits
C. In microwave circuits
D. In very low frequency RF circuits

E6A02 (A)
Which of the following semiconductor materials contains excess free electrons?
A. N-type
B. P-type
C. Bipolar
D. Insulated gate

E6A03 (C)
Why does a PN-junction diode not conduct current when reverse biased?
A. Only P-type semiconductor material can conduct current
B. Only N-type semiconductor material can conduct current
C. Holes in P-type material and electrons in the N-type material are separated by the applied voltage, widening the depletion region
D. Excess holes in P-type material combine with the electrons in N-type material, converting the entire diode into an insulator

E6A04 (C)
What is the name given to an impurity atom that adds holes to a semiconductor crystal structure?
A. Insulator impurity
B. N-type impurity
C. Acceptor impurity
D. Donor impurity

E6A05 (C)
What is the alpha of a bipolar junction transistor?
A. The change of collector current with respect to base current
B. The change of base current with respect to collector current
C. The change of collector current with respect to emitter current
D. The change of collector current with respect to gate current

E6A06 (B)
What is the beta of a bipolar junction transistor?
A. The frequency at which the current gain is reduced to 1
B. The change in collector current with respect to base current
C. The breakdown voltage of the base to collector junction
D. The switching speed of the transistor

E6A07 (D)
Which of the following indicates that a silicon NPN junction transistor is biased on?
A. Base-to-emitter resistance of approximately 6 to 7 ohms
B. Base-to-emitter resistance of approximately 0.6 to 0.7 ohms
C. Base-to-emitter voltage of approximately 6 to 7 volts
D. Base-to-emitter voltage of approximately 0.6 to 0.7 volts

E6A08 (D)
What term indicates the frequency at which the grounded-base current gain of a transistor has decreased to 0.7 of the gain obtainable at 1 kHz?
A. Corner frequency
B. Alpha rejection frequency
C. Beta cutoff frequency
D. Alpha cutoff frequency

E6A09 (A)
What is a depletion-mode FET?
A. An FET that exhibits a current flow between source and drain when no gate voltage is applied
B. An FET that has no current flow between source and drain when no gate voltage is applied
C. Any FET without a channel
D. Any FET for which holes are the majority carriers

E6A10 (B)
In Figure E6-2, what is the schematic symbol for an N-channel dual-gate MOSFET?
A. 2
B. 4
C. 5
D. 6

E6A11 (A)
In Figure E6-2, what is the schematic symbol for a P-channel junction FET?
A. 1
B. 2
C. 3
D. 6

E6A12 (D)
Why do many MOSFET devices have internally connected Zener diodes on the gates?
A. To provide a voltage reference for the correct amount of reverse-bias gate voltage
B. To protect the substrate from excessive voltages
C. To keep the gate voltage within specifications and prevent the device from overheating
D. To reduce the chance of the gate insulation being punctured by static discharges or excessive voltages

E6A13 (C)
What do the initials CMOS stand for?
A. Common Mode Oscillating System
B. Complementary Mica-Oxide Silicon
C. Complementary Metal-Oxide Semiconductor
D. Common Mode Organic Silicon

E6A14 (C)
How does DC input impedance at the gate of a field-effect transistor compare with the DC input impedance of a bipolar transistor?
A. They are both low impedance
B. An FET has low input impedance; a bipolar transistor has high input impedance
C. An FET has high input impedance; a bipolar transistor has low input impedance
D. They are both high impedance

E6A15 (B)
Which semiconductor material contains excess holes in the outer shell of electrons?
A. N-type
B. P-type
C. Superconductor-type
D. Bipolar-type

E6A16 (B)
What are the majority charge carriers in N-type semiconductor material?
A. Holes
B. Free electrons
C. Free protons
D. Free neutrons

E6A17 (D)
What are the names of the three terminals of a field-effect transistor?
A. Gate 1, gate 2, drain
B. Emitter, base, collector
C. Emitter, base 1, base 2
D. Gate, drain, source

E6B - DIODES

E6B01 (B)
What is the most useful characteristic of a Zener diode?
A. A constant current drop under conditions of varying voltage
B. A constant voltage drop under conditions of varying current
C. A negative resistance region
D. An internal capacitance that varies with the applied voltage

E6B02 (D)
What is an important characteristic of a Schottky diode as compared to an ordinary silicon diode when used as a power supply rectifier?
A. Much higher reverse voltage breakdown
B. Controlled reverse avalanche voltage
C. Enhanced carrier retention time
D. Less forward voltage drop

E6B03 (C)
What special type of diode is capable of both amplification and oscillation?
A. Point contact
B. Zener
C. Tunnel
D. Junction

E6B04 (A)
What type of semiconductor device is designed for use as a voltage-controlled capacitor?
A. Varactor diode
B. Tunnel diode
C. Silicon-controlled rectifier
D. Zener diode

E6B05 (D)
What characteristic of a PIN diode makes it useful as an RF switch or attenuator?
A. Extremely high reverse breakdown voltage
B. Ability to dissipate large amounts of power
C. Reverse bias controls its forward voltage drop
D. A large region of intrinsic material

E6B06 (D)
Which of the following is a common use of a hot-carrier diode?
A. As balanced mixers in FM generation
B. As a variable capacitance in an automatic frequency control circuit
C. As a constant voltage reference in a power supply
D. As a VHF/UHF mixer or detector

E6B07 (B)
What is the failure mechanism when a junction diode fails due to excessive current?
A. Excessive inverse voltage
B. Excessive junction temperature
C. Insufficient forward voltage
D. Charge carrier depletion

E6B08 (A)
Which of the following describes a type of semiconductor diode?
A. Metal-semiconductor junction
B. Electrolytic rectifier
C. CMOS-field effect
D. Thermionic emission diode

E6B09 (C)
What is a common use for point contact diodes?
A. As a constant current source
B. As a constant voltage source
C. As an RF detector
D. As a high voltage rectifier

E6B10 (B)
In Figure E6-3, what is the schematic symbol for a light-emitting diode?
A. 1
B. 5
C. 6
D. 7

E6B11 (A)
What is used to control the attenuation of RF signals by a PIN diode?
A. Forward DC bias current
B. A sub-harmonic pump signal
C. Reverse voltage larger than the RF signal
D. Capacitance of an RF coupling capacitor

E6B12 (C)
What is one common use for PIN diodes?
A. As a constant current source
B. As a constant voltage source
C. As an RF switch
D. As a high voltage rectifier

E6B13 (B)
What type of bias is required for an LED to emit light?
A. Reverse bias
B. Forward bias
C. Zero bias
D. Inductive bias

E6C - DIGITAL ICS: FAMILIES OF DIGITAL ICS; GATES; PROGRAMMABLE LOGIC DEVICES
   (PLDS)

E6C01 (A)
What is the function of hysteresis in a comparator?
A. To prevent input noise from causing unstable output signals
B. To allow the comparator to be used with AC input signal
C. To cause the output to change states continually
D. To increase the sensitivity

E6C02 (B)
What happens when the level of a comparator's input signal crosses the threshold?
A. The IC input can be damaged
B. The comparator changes its output state
C. The comparator enters latch-up
D. The feedback loop becomes unstable

E6C03 (A)
What is tri-state logic?
A. Logic devices with 0, 1, and high impedance output states
B. Logic devices that utilize ternary math
C. Low power logic devices designed to operate at 3 volts
D. Proprietary logic devices manufactured by Tri-State Devices

E6C04 (B)
What is the primary advantage of tri-state logic?
A. Low power consumption
B. Ability to connect many device outputs to a common bus
C. High speed operation
D. More efficient arithmetic operations

E6C05 (D)
What is an advantage of CMOS logic devices over TTL devices?
A. Differential output capability
B. Lower distortion
C. Immune to damage from static discharge
D. Lower power consumption

E6C06 (C)
Why do CMOS digital integrated circuits have high immunity to noise on the input signal or power supply?
A. Larger bypass capacitors are used in CMOS circuit design
B. The input switching threshold is about two times the power supply voltage
C. The input switching threshold is about one-half the power supply voltage
D. Input signals are stronger

E6C07 (B)
What best describes a pull-up or pull-down resistor?
A. A resistor in a keying circuit used to reduce key clicks
B. A resistor connected to the positive or negative supply line used to establish a voltage when an input or output is an open circuit
C. A resistor that insures that an oscillator frequency does not drive lower over time
D. A resistor connected to an op-amp output that only functions when the logic output is false

E6C08 (B)
In Figure E6-5, what is the schematic symbol for a NAND gate?
A. 1
B. 2
C. 3
D. 4

E6C09 (B)
What is a Programmable Logic Device (PLD)?
A. A device to control industrial equipment
B. A programmable collection of logic gates and circuits in a single integrated circuit
C. Programmable equipment used for testing digital logic integrated circuits
D. An algorithm for simulating logic functions during circuit design

E6C10 (D)
In Figure E6-5, what is the schematic symbol for a NOR gate?
A. 1
B. 2
C. 3
D. 4

E6C11 (C)
In Figure E6-5, what is the schematic symbol for the NOT operation (inverter)?
A. 2
B. 4
C. 5
D. 6

E6C12 (D)
What is BiCMOS logic?
A. A logic device with two CMOS circuits per package
B. A FET logic family based on bimetallic semiconductors
C. A logic family based on bismuth CMOS devices
D. An integrated circuit logic family using both bipolar and CMOS transistors

E6C13 (C)
Which of the following is an advantage of BiCMOS logic?
A. Its simplicity results in much less expensive devices than standard CMOS
B. It is totally immune to electrostatic damage
C. It has the high input impedance of CMOS and the low output impedance of bipolar transistors
D. All of these choices are correct

E6C14 (B)
What is the primary advantage of using a Programmable Gate Array (PGA) in a logic circuit?
A. Many similar gates are less expensive than a mixture of gate types
B. Complex logic functions can be created in a single integrated circuit
C. A PGA contains its own internal power supply
D. All of these choices are correct

E6D - TOROIDAL AND SOLENOIDAL INDUCTORS: PERMEABILITY, CORE MATERIAL, SELECTING, WINDING; TRANSFORMERS; PIEZOELECTRIC DEVICES

E6D01 (A)
How many turns will be required to produce a 5-microhenry inductor using a powdered-iron toroidal core that has an inductance index (A L) value of 40 microhenrys/100 turns?
A. 35 turns
B. 13 turns
C. 79 turns
D. 141 turns

E6D02 (A)
What is the equivalent circuit of a quartz crystal?
A. Motional capacitance, motional inductance, and loss resistance in series, all in parallel with a shunt capacitor representing electrode and stray capacitance
B. Motional capacitance, motional inductance, loss resistance, and a capacitor representing electrode and stray capacitance all in parallel
C. Motional capacitance, motional inductance, loss resistance, and a capacitor representing electrode and stray capacitance all in series
D. Motional inductance and loss resistance in series, paralleled with motional capacitance and a capacitor representing electrode and stray capacitance

E6D03 (A)
Which of the following is an aspect of the piezoelectric effect?
A. Mechanical deformation of material by the application of a voltage
B. Mechanical deformation of material by the application of a magnetic field
C. Generation of electrical energy in the presence of light
D. Increased conductivity in the presence of light

E6D04 (B)
Which materials are commonly used as a slug core in a variable inductor?
A. Polystyrene and polyethylene
B. Ferrite and brass
C. Teflon and Delrin
D. Cobalt and aluminum

E6D05 (C)
What is one reason for using ferrite cores rather than powdered-iron in an inductor?
A. Ferrite toroids generally have lower initial permeability
B. Ferrite toroids generally have better temperature stability
C. Ferrite toroids generally require fewer turns to produce a given inductance value
D. Ferrite toroids are easier to use with surface mount technology

E6D06 (D)
What core material property determines the inductance of a toroidal inductor?
A. Thermal impedance
B. Resistance
C. Reactivity
D. Permeability

E6D07 (B)
What is the usable frequency range of inductors that use toroidal cores, assuming a correct selection of core material for the frequency being used?
A. From a few kHz to no more than 30 MHz
B. From less than 20 Hz to approximately 300 MHz
C. From approximately 10 Hz to no more than 3000 kHz
D. From about 100 kHz to at least 1000 GHz

E6D08 (B)
What is one reason for using powdered-iron cores rather than ferrite cores in an inductor?
A. Powdered-iron cores generally have greater initial permeability
B. Powdered-iron cores generally maintain their characteristics at higher currents
C. Powdered-iron cores generally require fewer turns to produce a given inductance
D. Powdered-iron cores use smaller diameter wire for the same inductance

E6D09 (C)
What devices are commonly used as VHF and UHF parasitic suppressors at the input and output terminals of a transistor HF amplifier?
A. Electrolytic capacitors
B. Butterworth filters
C. Ferrite beads
D. Steel-core toroids

E6D10 (A)
What is a primary advantage of using a toroidal core instead of a solenoidal core in an inductor?
A. Toroidal cores confine most of the magnetic field within the core material
B. Toroidal cores make it easier to couple the magnetic energy into other components
C. Toroidal cores exhibit greater hysteresis
D. Toroidal cores have lower Q characteristics

E6D11 (C)
How many turns will be required to produce a 1-mH inductor using a core that has an inductance index (A L) value of 523 millihenrys/1000 turns?
A. 2 turns
B. 4 turns
C. 43 turns
D. 229 turns

E6D12 (C)
What is the definition of saturation in a ferrite core inductor?
A. The inductor windings are over coupled
B. The inductor's voltage rating is exceeded causing a flashover
C. The ability of the inductor's core to store magnetic energy has been exceeded
D. Adjacent inductors become over-coupled

E6D13 (A)
What is the primary cause of inductor self-resonance?
A. Inter-turn capacitance
B. The skin effect
C. Inductive kickback
D. Non-linear core hysteresis

E6D14 (B)
Which type of slug material decreases inductance when inserted into a coil?
A. Ceramic
B. Brass
C. Ferrite
D. Powdered-iron

E6D15 (A)
What is current in the primary winding of a transformer called if no load is attached to the secondary?
A. Magnetizing current
B. Direct current
C. Excitation current
D. Stabilizing current

E6D16 (D)
What is the common name for a capacitor connected across a transformer secondary that is used to absorb transient voltage spikes?
A. Clipper capacitor
B. Trimmer capacitor
C. Feedback capacitor
D. Snubber capacitor

E6D17 (A)
Why should core saturation of a conventional impedance matching transformer be avoided?
A. Harmonics and distortion could result
B. Magnetic flux would increase with frequency
C. RF susceptance would increase
D. Temporary changes of the core permeability could result

E6E - ANALOG ICS: MMICS, CCDS, DEVICE PACKAGES

E6E01 (C)
Which of the following is true of a charge-coupled device (CCD)?
A. Its phase shift changes rapidly with frequency
B. It is a CMOS analog-to-digital converter
C. It samples an analog signal and passes it in stages from the input to the output
D. It is used in a battery charger circuit

E6E02 (A)
Which of the following device packages is a through-hole type?
A. DIP
B. PLCC
C. Ball grid array
D. SOT

E6E03 (D)
Which of the following materials is likely to provide the highest frequency of operation when used in MMICs?
A. Silicon
B. Silicon nitride
C. Silicon dioxide
D. Gallium nitride

E6E04 (A)
Which is the most common input and output impedance of circuits that use MMICs?
A. 50 ohms
B. 300 ohms
C. 450 ohms
D. 10 ohms

E6E05 (A)
Which of the following noise figure values is typical of a low-noise UHF preamplifier?
A. 2 dB
B. -10 dB
C. 44 dBm
D. -20 dBm

E6E06 (D)
What characteristics of the MMIC make it a popular choice for VHF through microwave circuits?
A. The ability to retrieve information from a single signal even in the presence of other strong signals
B. Plate current that is controlled by a control grid
C. Nearly infinite gain, very high input impedance, and very low output impedance
D. Controlled gain, low noise figure, and constant input and output impedance over the specified frequency range

E6E07 (B)
Which of the following is typically used to construct a MMIC-based microwave amplifier?
A. Ground-plane construction
B. Microstrip construction
C. Point-to-point construction
D. Wave-soldering construction

E6E08 (A)
How is voltage from a power supply normally furnished to the most common type of monolithic microwave integrated circuit (MMIC)?
A. Through a resistor and/or RF choke connected to the amplifier output lead
B. MMICs require no operating bias
C. Through a capacitor and RF choke connected to the amplifier input lead
D. Directly to the bias voltage (VCC IN) lead

E6E09 (D)
Which of the following component package types would be most suitable for use at frequencies above the HF range?
A. TO-220
B. Axial lead
C. Radial lead
D. Surface mount

E6E10 (D)
What is the packaging technique in which leadless components are soldered directly to circuit boards?
A. Direct soldering
B. Virtual lead mounting
C. Stripped lead
D. Surface mount

E6E11 (D)
What is a characteristic of DIP packaging used for integrated circuits?
A. Package mounts in a direct inverted position
B. Low leakage doubly insulated package
C. Two chips in each package (Dual In Package)
D. A total of two rows of connecting pins placed on opposite sides of the package (Dual In-line Package)

E6E12 (B)
Why are high-power RF amplifier ICs and transistors sometimes mounted in ceramic packages?
A. High-voltage insulating ability
B. Better dissipation of heat
C. Enhanced sensitivity to light
D. To provide a low-pass frequency response

E6F - OPTICAL COMPONENTS: PHOTOCONDUCTIVE PRINCIPLES AND EFFECTS, PHOTOVOLTAIC SYSTEMS, OPTICAL COUPLERS, OPTICAL SENSORS, AND OPTOISOLATORS; LCDS

E6F01 (B)
What is photoconductivity?
A. The conversion of photon energy to electromotive energy
B. The increased conductivity of an illuminated semiconductor
C. The conversion of electromotive energy to photon energy
D. The decreased conductivity of an illuminated semiconductor

E6F02 (A)
What happens to the conductivity of a photoconductive material when light shines on it?
A. It increases
B. It decreases
C. It stays the same
D. It becomes unstable

E6F03 (D)
What is the most common configuration of an optoisolator or optocoupler?
A. A lens and a photomultiplier
B. A frequency modulated helium-neon laser
C. An amplitude modulated helium-neon laser
D. An LED and a phototransistor

E6F04 (B)
What is the photovoltaic effect?
A. The conversion of voltage to current when exposed to light
B. The conversion of light to electrical energy
C. The conversion of electrical energy to mechanical energy
D. The tendency of a battery to discharge when used outside

E6F05 (A)
Which describes an optical shaft encoder?
A. A device which detects rotation of a control by interrupting a light source with a patterned wheel
B. A device which measures the strength of a beam of light using analog to digital conversion
C. A digital encryption device often used to encrypt spacecraft control signals
D. A device for generating RTTY signals by means of a rotating light source

E6F06 (A)
Which of these materials is affected the most by photoconductivity?
A. A crystalline semiconductor
B. An ordinary metal
C. A heavy metal
D. A liquid semiconductor

E6F07 (B)
What is a solid state relay?
A. A relay using transistors to drive the relay coil
B. A device that uses semiconductors to implement the functions of an electromechanical relay
C. A mechanical relay that latches in the on or off state each time it is pulsed
D. A passive delay line

E6F08 (C)
Why are optoisolators often used in conjunction with solid state circuits when switching 120VAC?
A. Optoisolators provide a low impedance link between a control circuit and a power circuit
B. Optoisolators provide impedance matching between the control circuit and power circuit
C. Optoisolators provide a very high degree of electrical isolation between a control circuit and the circuit being switched
D. Optoisolators eliminate the effects of reflected light in the control circuit

E6F09 (D)
What is the efficiency of a photovoltaic cell?
A. The output RF power divided by the input DC power
B. The effective payback period
C. The open-circuit voltage divided by the short-circuit current under full illumination
D. The relative fraction of light that is converted to current

E6F10 (B)
What is the most common type of photovoltaic cell used for electrical power generation?
A. Selenium
B. Silicon
C. Cadmium Sulfide
D. Copper oxide

E6F11 (B)
What is the approximate open-circuit voltage produced by a fully-illuminated silicon photovoltaic cell?
A. 0.1 V
B. 0.5 V
C. 1.5 V
D. 12 V

E6F12 (C)
What absorbs the energy from light falling on a photovoltaic cell?
A. Protons
B. Photons
C. Electrons
D. Holes

E6F13 (B)
What is a liquid crystal display (LCD)?
A. A modern replacement for a quartz crystal oscillator which displays its fundamental frequency
B. A display utilizing a crystalline liquid and polarizing filters which becomes opaque when voltage is applied
C. A frequency-determining unit for a transmitter or receiver
D. A display that uses a glowing liquid to remain brightly lit in dim light

E6F14 (B)
Which of the following is true of LCD displays?
A. They are hard to view in high ambient light conditions
B. They may be hard view through polarized lenses
C. They only display alphanumeric symbols
D. All of these choices are correct

Subelement E7 - Practical Circuits [8 Groups]

E7A - DIGITAL CIRCUITS: DIGITAL CIRCUIT PRINCIPLES AND LOGIC CIRCUITS: CLASSES OF LOGIC ELEMENTS; POSITIVE AND NEGATIVE LOGIC; FREQUENCY DIVIDERS; TRUTH TABLES

E7A01 (C)
Which is a bi-stable circuit?
A. An "AND" gate
B. An "OR" gate
C. A flip-flop
D. A clock

E7A02 (A)
What is the function of a decade counter digital IC?
A. It produces one output pulse for every ten input pulses
B. It decodes a decimal number for display on a seven segment LED display
C. It produces ten output pulses for every input pulse
D. It adds two decimal numbers together

E7A03 (B)
Which of the following can divide the frequency of a pulse train by 2?
A. An XOR gate
B. A flip-flop
C. An OR gate
D. A multiplexer

E7A04 (B)
How many flip-flops are required to divide a signal frequency by 4?
A. 1
B. 2
C. 4
D. 8

E7A05 (D)
Which of the following is a circuit that continuously alternates between two states without an external clock?
A. Monostable multivibrator
B. J-K flip-flop
C. T flip-flop
D. Astable multivibrator

E7A06 (A)
What is a characteristic of a monostable multivibrator?
A. It switches momentarily to the opposite binary state and then returns to its original state after a set time
B. It produces a continuous square wave oscillating between 1 and 0
C. It stores one bit of data in either a 0 or 1 state
D. It maintains a constant output voltage, regardless of variations in the input voltage

E7A07 (D)
What logical operation does a NAND gate perform?
A. It produces logic "0" at its output only when all inputs are logic "0"
B. It produces logic "1" at its output only when all inputs are logic "1"
C. It produces logic "0" at its output if some but not all inputs are logic "1"
D. It produces logic "0" at its output only when all inputs are logic "1"

E7A08 (A)
What logical operation does an OR gate perform?
A. It produces logic "1" at its output if any or all inputs are logic "1"
B. It produces logic "0" at its output if all inputs are logic "1"
C. It only produces logic "0" at its output when all inputs are logic "1"
D. It produces logic "1" at its output if all inputs are logic "0"

E7A09 (C)
What logical operation is performed by an exclusive NOR gate?
A. It produces logic "0" at its output only if all inputs are logic "0"
B. It produces logic "1" at its output only if all inputs are logic "1"
C. It produces logic "0" at its output if any single input is logic "1"
D. It produces logic "1" at its output if any single input is logic "1"

E7A10 (C)
What is a truth table?
A. A table of logic symbols that indicate the high logic states of an op-amp
B. A diagram showing logic states when the digital device output is true
C. A list of inputs and corresponding outputs for a digital device
D. A table of logic symbols that indicate the logic states of an op-amp

E7A11 (D)
What type of logic defines "1" as a high voltage?
A. Reverse Logic
B. Assertive Logic
C. Negative logic
D. Positive Logic

E7A12 (C)
What type of logic defines "0" as a high voltage?
A. Reverse Logic
B. Assertive Logic
C. Negative logic
D. Positive Logic

E7B - AMPLIFIERS: CLASS OF OPERATION; VACUUM TUBE AND SOLID-STATE CIRCUITS;
  DISTORTION AND INTERMODULATION; SPURIOUS AND PARASITIC SUPPRESSION;
  MICROWAVE AMPLIFIERS; SWITCHING-TYPE AMPLIFIERS

E7B01 (A)
For what portion of a signal cycle does a Class AB amplifier operate?
A. More than 180 degrees but less than 360 degrees
B. Exactly 180 degrees
C. The entire cycle
D. Less than 180 degrees

E7B02 (A)
What is a Class D amplifier?
A. A type of amplifier that uses switching technology to achieve high efficiency
B. A low power amplifier that uses a differential amplifier for improved linearity
C. An amplifier that uses drift-mode FETs for high efficiency
D. A frequency doubling amplifier

E7B03 (A)
Which of the following components form the output of a class D amplifier circuit?
A. A low-pass filter to remove switching signal components
B. A high-pass filter to compensate for low gain at low frequencies
C. A matched load resistor to prevent damage by switching transients
D. A temperature compensating load resistor to improve linearity

E7B04 (A)
Where on the load line of a Class A common emitter amplifier would bias normally be set?
A. Approximately half-way between saturation and cutoff
B. Where the load line intersects the voltage axis
C. At a point where the bias resistor equals the load resistor
D. At a point where the load line intersects the zero bias current curve

E7B05 (C)
What can be done to prevent unwanted oscillations in an RF power amplifier?
A. Tune the stage for maximum SWR
B. Tune both the input and output for maximum power
C. Install parasitic suppressors and/or neutralize the stage
D. Use a phase inverter in the output filter

E7B06 (B)
Which of the following amplifier types reduces or eliminates even order harmonics?
A. Push-push
B. Push-pull
C. Class C
D. Class AB

E7B07 (D)
Which of the following is a likely result when a Class C amplifier is used to amplify a single-sideband phone signal?
A. Reduced intermodulation products
B. Increased overall intelligibility
C. Signal inversion
D. Signal distortion and excessive bandwidth

E7B08 (C)
How can an RF power amplifier be neutralized?
A. By increasing the driving power
B. By reducing the driving power
C. By feeding a 180-degree out-of-phase portion of the output back to the input
D. By feeding an in-phase component of the output back to the input

E7B09 (D)
Which of the following describes how the loading and tuning capacitors are to be adjusted when tuning a vacuum tube RF power amplifier that employs a Pi-network output circuit?
A. The loading capacitor is set to maximum capacitance and the tuning capacitor is adjusted for minimum allowable plate current
B. The tuning capacitor is set to maximum capacitance and the loading capacitor is adjusted for minimum plate permissible current
C. The loading capacitor is adjusted to minimum plate current while alternately adjusting the tuning capacitor for maximum allowable plate current
D. The tuning capacitor is adjusted for minimum plate current, and the loading capacitor is adjusted for maximum permissible plate current

E7B10 (B)
In Figure E7-1, what is the purpose of R1 and R2?
A. Load resistors
B. Fixed bias
C. Self bias
D. Feedback

E7B11 (D)
In Figure E7-1, what is the purpose of R3?
A. Fixed bias
B. Emitter bypass
C. Output load resistor
D. Self bias

E7B12 (C)
What type of amplifier circuit is shown in Figure E7-1?
A. Common base
B. Common collector
C. Common emitter
D. Emitter follower

E7B13 (A)
In Figure E7-2, what is the purpose of R?
A. Emitter load
B. Fixed bias
C. Collector load
D. Voltage regulation

E7B14 (B)
Why are switching amplifiers more efficient than linear amplifiers?
A. Switching amplifiers operate at higher voltages
B. The power transistor is at saturation or cut off most of the time, resulting in low power dissipation
C. Linear amplifiers have high gain resulting in higher harmonic content
D. Switching amplifiers use push-pull circuits

E7B15 (C)
What is one way to prevent thermal runaway in a bipolar transistor amplifier?
A. Neutralization
B. Select transistors with high beta
C. Use a resistor in series with the emitter
D. All of these choices are correct

E7B16 (A)
What is the effect of intermodulation products in a linear power amplifier?
A. Transmission of spurious signals
B. Creation of parasitic oscillations
C. Low efficiency
D. All of these choices are correct

E7B17 (A)
Why are odd-order rather than even-order intermodulation distortion products of concern in linear power amplifiers?
A. Because they are relatively close in frequency to the desired signal
B. Because they are relatively far in frequency from the desired signal
C. Because they invert the sidebands causing distortion
D. Because they maintain the sidebands, thus causing multiple duplicate signals

E7B18 (C)
What is a characteristic of a grounded-grid amplifier?
A. High power gain
B. High filament voltage
C. Low input impedance
D. Low bandwidth

E7C01 (D)
How are the capacitors and inductors of a low-pass filter Pi-network arranged between the network's input and output?
A. Two inductors are in series between the input and output, and a capacitor is connected between the two inductors and ground
B. Two capacitors are in series between the input and output, and an inductor is connected between the two capacitors and ground
C. An inductor is connected between the input and ground, another inductor is connected between the output and ground, and a capacitor is connected between the input and output
D. A capacitor is connected between the input and ground, another capacitor is connected between the output and ground, and an inductor is connected between input and output

E7C02 (C)
Which of the following is a property of a T-network with series capacitors and a parallel shunt inductor?
A. It is a low-pass filter
B. It is a band-pass filter
C. It is a high-pass filter
D. It is a notch filter

E7C03 (A)
What advantage does a Pi-L-network have over a regular Pi-network for impedance matching between the final amplifier of a vacuum-tube transmitter and an antenna?
A. Greater harmonic suppression
B. Higher efficiency
C. Lower losses
D. Greater transformation range

E7C04 (C)
How does an impedance-matching circuit transform a complex impedance to a resistive impedance?
A. It introduces negative resistance to cancel the resistive part of impedance
B. It introduces transconductance to cancel the reactive part of impedance
C. It cancels the reactive part of the impedance and changes the resistive part to a desired value
D. Network resistances are substituted for load resistances and reactances are matched to the resistances

E7C05 (D)
Which filter type is described as having ripple in the passband and a sharp cutoff?
A. A Butterworth filter
B. An active LC filter
C. A passive op-amp filter
D. A Chebyshev filter

E7C06 (C)
What are the distinguishing features of an elliptical filter?
A. Gradual passband rolloff with minimal stop band ripple
B. Extremely flat response over its pass band with gradually rounded stop band corners
C. Extremely sharp cutoff with one or more notches in the stop band
D. Gradual passband rolloff with extreme stop band ripple

E7C07 (B)
What kind of filter would you use to attenuate an interfering carrier signal while receiving an SSB transmission?
A. A band-pass filter
B. A notch filter
C. A Pi-network filter
D. An all-pass filter

E7C08 (A)
Which of the following factors has the greatest effect in helping determine the bandwidth and response shape of a crystal ladder filter?
A. The relative frequencies of the individual crystals
B. The DC voltage applied to the quartz crystal
C. The gain of the RF stage preceding the filter
D. The amplitude of the signals passing through the filter

E7C09 (B)
What is a Jones filter as used as part of an HF receiver IF stage?
A. An automatic notch filter
B. A variable bandwidth crystal lattice filter
C. A special filter that emphasizes image responses
D. A filter that removes impulse noise

E7C10 (B)
Which of the following filters would be the best choice for use in a 2 meter repeater duplexer?
A. A crystal filter
B. A cavity filter
C. A DSP filter
D. An L-C filter

E7C11 (D)
Which of the following is the common name for a filter network which is equivalent to two L-networks connected back-to-back with the two inductors in series and the capacitors in shunt at the input and output?
A. Pi-L
B. Cascode
C. Omega
D. Pi

E7C12 (B)
Which describes a Pi-L-network used for matching a vacuum tube final amplifier to a 50 ohm unbalanced output?
A. A Phase Inverter Load network
B. A Pi-network with an additional series inductor on the output
C. A network with only three discrete parts
D. A matching network in which all components are isolated from ground

E7C13 (A)
What is one advantage of a Pi-matching network over an L-matching network consisting of a single inductor and a single capacitor?
A. The Q of Pi-networks can be varied depending on the component values chosen
B. L-networks cannot perform impedance transformation
C. Pi-networks have fewer components
D. Pi-networks are designed for balanced input and output

E7C14 (C)
Which mode is most affected by non-linear phase response in a receiver IF filter?
A. Meteor scatter
B. Single-Sideband voice
C. Digital
D. Video

E7C15 (D)
What is a crystal lattice filter?
A. A power supply filter made with interlaced quartz crystals
B. An audio filter made with four quartz crystals that resonate at 1kHz intervals
C. A filter with wide bandwidth and shallow skirts made using quartz crystals
D. A filter with narrow bandwidth and steep skirts made using quartz crystals

E7D - POWER SUPPLIES AND VOLTAGE REGULATORS; SOLAR ARRAY CHARGE
  CONTROLLERS

E7D01 (D)
What is one characteristic of a linear electronic voltage regulator?
A. It has a ramp voltage as its output
B. It eliminates the need for a pass transistor
C. The control element duty cycle is proportional to the line or load conditions
D. The conduction of a control element is varied to maintain a constant output
voltage

E7D02 (C)
What is one characteristic of a switching electronic voltage regulator?
A. The resistance of a control element is varied in direct proportion to the line
voltage or load current
B. It is generally less efficient than a linear regulator
C. The controlled device's duty cycle is changed to produce a constant average
output voltage
D. It gives a ramp voltage at its output

E7D03 (A)
What device is typically used as a stable reference voltage in a linear voltage
regulator?
A. A Zener diode
B. A tunnel diode
C. An SCR
D. A varactor diode

E7D04 (B)
Which of the following types of linear voltage regulator usually make the most
efficient use of the primary power source?
A. A series current source
B. A series regulator
C. A shunt regulator
D. A shunt current source

E7D05 (D)
Which of the following types of linear voltage regulator places a constant load on the
unregulated voltage source?
A. A constant current source
B. A series regulator
C. A shunt current source
D. A shunt regulator

E7D06 (C)
What is the purpose of Q1 in the circuit shown in Figure E7-3?
A. It provides negative feedback to improve regulation
B. It provides a constant load for the voltage source
C. It increases the current-handling capability of the regulator
D. It provides D1 with current

E7D07 (A)
What is the purpose of C2 in the circuit shown in Figure E7-3?
A. It bypasses hum around D1
B. It is a brute force filter for the output
C. To self-resonate at the hum frequency
D. To provide fixed DC bias for Q1

E7D08 (C)
What type of circuit is shown in Figure E7-3?
A. Switching voltage regulator
B. Grounded emitter amplifier
C. Linear voltage regulator
D. Emitter follower

E7D09 (C)
What is the main reason to use a charge controller with a solar power system?
A. Prevention of battery undercharge
B. Control of electrolyte levels during battery discharge
C. Prevention of battery damage due to overcharge
D. Matching of day and night charge rates

E7D10 (C)
What is the primary reason that a high-frequency switching type high voltage power supply can be both less expensive and lighter in weight than a conventional power supply?
A. The inverter design does not require any output filtering
B. It uses a diode bridge rectifier for increased output
C. The high frequency inverter design uses much smaller transformers and filter components for an equivalent power output
D. It uses a large power factor compensation capacitor to create free power from the unused portion of the AC cycle

E7D11 (D)
What circuit element is controlled by a series analog voltage regulator to maintain a constant output voltage?
A. Reference voltage
B. Switching inductance
C. Error amplifier
D. Pass transistor

E7D12 (C)
What is the drop-out voltage of an analog voltage regulator?
A. Minimum input voltage for rated power dissipation
B. Maximum amount that the output voltage drops when the input voltage is varied over its specified range
C. Minimum input-to-output voltage required to maintain regulation
D. Maximum amount that the output voltage may decrease at rated load

E7D13 (C)
What is the equation for calculating power dissipation by a series connected linear voltage regulator?
A. Input voltage multiplied by input current
B. Input voltage divided by output current
C. Voltage difference from input to output multiplied by output current
D. Output voltage multiplied by output current

E7D14 (C)
What is one purpose of a "bleeder" resistor in a conventional unregulated power supply?
A. To cut down on waste heat generated by the power supply
B. To balance the low-voltage filament windings
C. To improve output voltage regulation
D. To boost the amount of output current

E7D15 (D)
What is the purpose of a "step-start" circuit in a high voltage power supply?
A. To provide a dual-voltage output for reduced power applications
B. To compensate for variations of the incoming line voltage
C. To allow for remote control of the power supply
D. To allow the filter capacitors to charge gradually

E7D16 (D)
When several electrolytic filter capacitors are connected in series to increase the operating voltage of a power supply filter circuit, why should resistors be connected across each capacitor?
A. To equalize, as much as possible, the voltage drop across each capacitor
B. To provide a safety bleeder to discharge the capacitors when the supply is off
C. To provide a minimum load current to reduce voltage excursions at light loads
D. All of these choices are correct

E7E - MODULATION AND DEMODULATION: REACTANCE, PHASE AND BALANCED MODULATORS; DETECTORS; MIXER STAGES

E7E01 (B)
Which of the following can be used to generate FM phone emissions?
A. A balanced modulator on the audio amplifier
B. A reactance modulator on the oscillator
C. A reactance modulator on the final amplifier
D. A balanced modulator on the oscillator

E7E02 (D)
What is the function of a reactance modulator?
A. To produce PM signals by using an electrically variable resistance
B. To produce AM signals by using an electrically variable inductance or capacitance
C. To produce AM signals by using an electrically variable resistance
D. To produce PM signals by using an electrically variable inductance or capacitance

E7E03 (C)
How does an analog phase modulator function?
A. By varying the tuning of a microphone preamplifier to produce PM signals
B. By varying the tuning of an amplifier tank circuit to produce AM signals
C. By varying the tuning of an amplifier tank circuit to produce PM signals
D. By varying the tuning of a microphone preamplifier to produce AM signals

E7E04 (A)
What is one way a single-sideband phone signal can be generated?
A. By using a balanced modulator followed by a filter
B. By using a reactance modulator followed by a mixer
C. By using a loop modulator followed by a mixer
D. By driving a product detector with a DSB signal

E7E05 (D)
What circuit is added to an FM transmitter to boost the higher audio frequencies?
A. A de-emphasis network
B. A heterodyne suppressor
C. An audio prescaler
D. A pre-emphasis network

E7E06 (A)
Why is de-emphasis commonly used in FM communications receivers?
A. For compatibility with transmitters using phase modulation
B. To reduce impulse noise reception
C. For higher efficiency
D. To remove third-order distortion products

E7E07 (B)
What is meant by the term baseband in radio communications?
A. The lowest frequency band that the transmitter or receiver covers
B. The frequency components present in the modulating signal
C. The unmodulated bandwidth of the transmitted signal
D. The basic oscillator frequency in an FM transmitter that is multiplied to increase the deviation and carrier frequency

E7E08 (C)
What are the principal frequencies that appear at the output of a mixer circuit?
A. Two and four times the original frequency
B. The sum, difference and square root of the input frequencies
C. The two input frequencies along with their sum and difference frequencies
D. 1.414 and 0.707 times the input frequency

E7E09 (A)
What occurs when an excessive amount of signal energy reaches a mixer circuit?
A. Spurious mixer products are generated
B. Mixer blanking occurs
C. Automatic limiting occurs
D. A beat frequency is generated

E7E10 (A)
How does a diode detector function?
A. By rectification and filtering of RF signals
B. By breakdown of the Zener voltage
C. By mixing signals with noise in the transition region of the diode
D. By sensing the change of reactance in the diode with respect to frequency

E7E11 (C)
Which type of detector is used for demodulating SSB signals?
A. Discriminator
B. Phase detector
C. Product detector
D. Phase comparator

E7E12 (D)
What is a frequency discriminator stage in a FM receiver?
A. An FM generator circuit
B. A circuit for filtering two closely adjacent signals
C. An automatic band-switching circuit
D. A circuit for detecting FM signals

E7F - DSP FILTERING AND OTHER OPERATIONS; SOFTWARE DEFINED RADIO
    FUNDAMENTALS; DSP MODULATION AND DEMODULATION

E7F01 (C)
What is meant by direct digital conversion as applied to software defined radios?
A. Software is converted from source code to object code during operation of the receiver
B. Incoming RF is converted to a control voltage for a voltage controlled oscillator
C. Incoming RF is digitized by an analog-to-digital converter without being mixed with a local oscillator signal
D. A switching mixer is used to generate I and Q signals directly from the RF input

E7F02 (A)
What kind of digital signal processing audio filter is used to remove unwanted noise from a received SSB signal?
A. An adaptive filter
B. A crystal-lattice filter
C. A Hilbert-transform filter
D. A phase-inverting filter

E7F03 (C)
What type of digital signal processing filter is used to generate an SSB signal?
A. An adaptive filter
B. A notch filter
C. A Hilbert-transform filter
D. An elliptical filter

E7F04 (D)
What is a common method of generating an SSB signal using digital signal processing?
A. Mixing products are converted to voltages and subtracted by adder circuits
B. A frequency synthesizer removes the unwanted sidebands
C. Emulation of quartz crystal filter characteristics
D. Combine signals with a quadrature phase relationship

E7F05 (B)
How frequently must an analog signal be sampled by an analog-to-digital converter so that the signal can be accurately reproduced?
A. At half the rate of the highest frequency component of the signal
B. At twice the rate of the highest frequency component of the signal
C. At the same rate as the highest frequency component of the signal
D. At four times the rate of the highest frequency component of the signal

E7F06 (D)
What is the minimum number of bits required for an analog-to-digital converter to sample a signal with a range of 1 volt at a resolution of 1 millivolt?
A. 4 bits
B. 6 bits
C. 8 bits
D. 10 bits

E7F07 (C)
What function can a Fast Fourier Transform perform?
A. Converting analog signals to digital form
B. Converting digital signals to analog form
C. Converting digital signals from the time domain to the frequency domain
D. Converting 8-bit data to 16 bit data

E7F08 (B)
What is the function of decimation with regard to digital filters?
A. Converting data to binary code decimal form
B. Reducing the effective sample rate by removing samples
C. Attenuating the signal
D. Removing unnecessary significant digits

E7F09 (A)
Why is an anti-aliasing digital filter required in a digital decimator?
A. It removes high-frequency signal components which would otherwise be reproduced as lower frequency components
B. It peaks the response of the decimator, improving bandwidth
C. It removes low frequency signal components to eliminate the need for DC restoration
D. It notches out the sampling frequency to avoid sampling errors

E7F10 (A)
What aspect of receiver analog-to-digital conversion determines the maximum receive bandwidth of a Direct Digital Conversion SDR?
A. Sample rate
B. Sample width in bits
C. Sample clock phase noise
D. Processor latency

E7F11 (B)
What sets the minimum detectable signal level for an SDR in the absence of atmospheric or thermal noise?
A. Sample clock phase noise
B. Reference voltage level and sample width in bits
C. Data storage transfer rate
D. Missing codes and jitter

E7F12 (A)
What digital process is applied to I and Q signals in order to recover the baseband modulation information?
A. Fast Fourier Transform
B. Decimation
C. Signal conditioning
D. Quadrature mixing

E7F13 (D)
What is the function of taps in a digital signal processing filter?
A. To reduce excess signal pressure levels
B. Provide access for debugging software
C. Select the point at which baseband signals are generated
D. Provide incremental signal delays for filter algorithms

E7F14 (B)
Which of the following would allow a digital signal processing filter to create a sharper filter response?
A. Higher data rate
B. More taps
C. Complex phasor representations
D. Double-precision math routines

E7F15 (A)
Which of the following is an advantage of a Finite Impulse Response (FIR) filter vs an Infinite Impulse Response (IIR) digital filter?
A. FIR filters delay all frequency components of the signal by the same amount
B. FIR filters are easier to implement for a given set of passband rolloff requirements
C. FIR filters can respond faster to impulses
D. All of these choices are correct

E7F16 (D)
How might the sampling rate of an existing digital signal be adjusted by a factor of 3/4?
A. Change the gain by a factor of 3/4
B. Multiply each sample value by a factor of 3/4
C. Add 3 to each input value and subtract 4 from each output value
D. Interpolate by a factor of three, then decimate by a factor of four

E7F17 (D)
What do the letters I and Q in I/Q Modulation represent?
A. Inactive and Quiescent
B. Instantaneous and Quasi-stable
C. Instantaneous and Quenched
D. In-phase and Quadrature

E7G - ACTIVE FILTERS AND OP-AMP CIRCUITS: ACTIVE AUDIO FILTERS; CHARACTERISTICS; BASIC CIRCUIT DESIGN; OPERATIONAL AMPLIFIERS

E7G01 (A)
What is the typical output impedance of an integrated circuit op-amp?
A. Very low
B. Very high
C. 100 ohms
D. 1000 ohms

E7G02 (D)
What is the effect of ringing in a filter?
A. An echo caused by a long time delay
B. A reduction in high frequency response
C. Partial cancellation of the signal over a range of frequencies
D. Undesired oscillations added to the desired signal

E7G03 (D)
What is the typical input impedance of an integrated circuit op-amp?
A. 100 ohms
B. 1000 ohms
C. Very low
D. Very high

E7G04 (C)
What is meant by the term op-amp input offset voltage?
A. The output voltage of the op-amp minus its input voltage
B. The difference between the output voltage of the op-amp and the input voltage required in the immediately following stage
C. The differential input voltage needed to bring the open loop output voltage to zero
D. The potential between the amplifier input terminals of the op-amp in an open loop condition

E7G05 (A)
How can unwanted ringing and audio instability be prevented in a multi-section op-amp RC audio filter circuit?
A. Restrict both gain and Q
B. Restrict gain but increase Q
C. Restrict Q but increase gain
D. Increase both gain and Q

E7G06 (D)
Which of the following is the most appropriate use of an op-amp active filter?
A. As a high-pass filter used to block RFI at the input to receivers
B. As a low-pass filter used between a transmitter and a transmission line
C. For smoothing power supply output
D. As an audio filter in a receiver

E7G07 (C)
What magnitude of voltage gain can be expected from the circuit in Figure E7-4 when R1 is 10 ohms and RF is 470 ohms?
A. 0.21
B. 94
C. 47
D. 24

E7G08 (D)
How does the gain of an ideal operational amplifier vary with frequency?
A. It increases linearly with increasing frequency
B. It decreases linearly with increasing frequency
C. It decreases logarithmically with increasing frequency
D. It does not vary with frequency

E7G09 (D)
What will be the output voltage of the circuit shown in Figure E7-4 if R1 is 1000 ohms, RF is 10,000 ohms, and 0.23 volts DC is applied to the input?
A. 0.23 volts
B. 2.3 volts
C. -0.23 volts
D. -2.3 volts

E7G10 (C)
What absolute voltage gain can be expected from the circuit in Figure E7-4 when R1 is 1800 ohms and RF is 68 kilohms?
A. 1
B. 0.03
C. 38
D. 76

E7G11 (B)
What absolute voltage gain can be expected from the circuit in Figure E7-4 when R1 is 3300 ohms and RF is 47 kilohms?
A. 28
B. 14
C. 7
D. 0.07

E7G12 (A)
What is an integrated circuit operational amplifier?
A. A high-gain, direct-coupled differential amplifier with very high input impedance and very low output impedance
B. A digital audio amplifier whose characteristics are determined by components external to the amplifier
C. An amplifier used to increase the average output of frequency modulated amateur signals to the legal limit
D. A RF amplifier used in the UHF and microwave regions

E7H01 (D)
What are three oscillator circuits used in Amateur Radio equipment?
A. Taft, Pierce and negative feedback
B. Pierce, Fenner and Beane
C. Taft, Hartley and Pierce
D. Colpitts, Hartley and Pierce

E7H02 (C)
Which describes a microphonic?
A. An IC used for amplifying microphone signals
B. Distortion caused by RF pickup on the microphone cable
C. Changes in oscillator frequency due to mechanical vibration
D. Excess loading of the microphone by an oscillator

E7H03 (A)
How is positive feedback supplied in a Hartley oscillator?
A. Through a tapped coil
B. Through a capacitive divider
C. Through link coupling
D. Through a neutralizing capacitor

E7H04 (C)
How is positive feedback supplied in a Colpitts oscillator?
A. Through a tapped coil
B. Through link coupling
C. Through a capacitive divider
D. Through a neutralizing capacitor

E7H05 (D)
How is positive feedback supplied in a Pierce oscillator?
A. Through a tapped coil
B. Through link coupling
C. Through a neutralizing capacitor
D. Through a quartz crystal

E7H06 (B)
Which of the following oscillator circuits are commonly used in VFOs?
A. Pierce and Zener
B. Colpitts and Hartley
C. Armstrong and deForest
D. Negative feedback and balanced feedback

E7H07 (D)
How can an oscillator's microphonic responses be reduced?
A. Use of NP0 capacitors
B. Eliminating noise on the oscillator's power supply
C. Using the oscillator only for CW and digital signals
D. Mechanically isolating the oscillator circuitry from its enclosure

E7H08 (A)
Which of the following components can be used to reduce thermal drift in crystal oscillators?
A. NP0 capacitors
B. Toroidal inductors
C. Wirewound resistors
D. Non-inductive resistors

E7H09 (A)
What type of frequency synthesizer circuit uses a phase accumulator, lookup table, digital to analog converter, and a low-pass anti-alias filter?
A. A direct digital synthesizer
B. A hybrid synthesizer
C. A phase locked loop synthesizer
D. A diode-switching matrix synthesizer

E7H10 (B)
What information is contained in the lookup table of a direct digital frequency synthesizer?
A. The phase relationship between a reference oscillator and the output waveform
B. The amplitude values that represent a sine-wave output
C. The phase relationship between a voltage-controlled oscillator and the output waveform
D. The synthesizer frequency limits and frequency values stored in the radio memories

E7H11 (C)
What are the major spectral impurity components of direct digital synthesizers?
A. Broadband noise
B. Digital conversion noise
C. Spurious signals at discrete frequencies
D. Nyquist limit noise

E7H12 (B)
Which of the following must be done to insure that a crystal oscillator provides the frequency specified by the crystal manufacturer?
A. Provide the crystal with a specified parallel inductance
B. Provide the crystal with a specified parallel capacitance
C. Bias the crystal at a specified voltage
D. Bias the crystal at a specified current

E7H13 (D)
Which of the following is a technique for providing highly accurate and stable oscillators needed for microwave transmission and reception?
A. Use a GPS signal reference
B. Use a rubidium stabilized reference oscillator
C. Use a temperature-controlled high Q dielectric resonator
D. All of these choices are correct

E7H14 (C)
What is a phase-locked loop circuit?
A. An electronic servo loop consisting of a ratio detector, reactance modulator, and voltage-controlled oscillator
B. An electronic circuit also known as a monostable multivibrator
C. An electronic servo loop consisting of a phase detector, a low-pass filter, a voltage-controlled oscillator, and a stable reference oscillator
D. An electronic circuit consisting of a precision push-pull amplifier with a differential input

E7H15 (D)
Which of these functions can be performed by a phase-locked loop?
A. Wide-band AF and RF power amplification
B. Comparison of two digital input signals, digital pulse counter
C. Photovoltaic conversion, optical coupling
D. Frequency synthesis, FM demodulation

Subelement E8 - Signals and Emissions [4 Groups]

E8A - AC WAVEFORMS: SINE, SQUARE, SAWTOOTH AND IRREGULAR WAVEFORMS; AC MEASUREMENTS; AVERAGE AND PEP OF RF SIGNALS; FOURIER ANALYSIS; ANALOG TO DIGITAL CONVERSION: DIGITAL TO ANALOG CONVERSION

E8A01 (A)
What is the name of the process that shows that a square wave is made up of a sine wave plus all of its odd harmonics?
A. Fourier analysis
B. Vector analysis
C. Numerical analysis
D. Differential analysis

E8A02 (C)
What type of wave has a rise time significantly faster than its fall time (or vice versa)?
A. A cosine wave
B. A square wave
C. A sawtooth wave
D. A sine wave

E8A03 (A)
What type of wave does a Fourier analysis show to be made up of sine waves of a given fundamental frequency plus all of its harmonics?
A. A sawtooth wave
B. A square wave
C. A sine wave
D. A cosine wave

E8A04 (B)
What is "dither" with respect to analog to digital converters?
A. An abnormal condition where the converter cannot settle on a value to represent the signal
B. A small amount of noise added to the input signal to allow more precise representation of a signal over time
C. An error caused by irregular quantization step size
D. A method of decimation by randomly skipping samples

E8A05 (D)
What would be the most accurate way of measuring the RMS voltage of a complex waveform?
A. By using a grid dip meter
B. By measuring the voltage with a D'Arsonval meter
C. By using an absorption wave meter
D. By measuring the heating effect in a known resistor

E8A06 (A)
What is the approximate ratio of PEP-to-average power in a typical single-sideband phone signal?
A. 2.5 to 1
B. 25 to 1
C. 1 to 1
D. 100 to 1

E8A07 (B)
What determines the PEP-to-average power ratio of a single-sideband phone signal?
A. The frequency of the modulating signal
B. The characteristics of the modulating signal
C. The degree of carrier suppression
D. The amplifier gain

E8A08 (C)
Why would a direct or flash conversion analog-to-digital converter be useful for a software defined radio?
A. Very low power consumption decreases frequency drift
B. Immunity to out of sequence coding reduces spurious responses
C. Very high speed allows digitizing high frequencies
D. All of these choices are correct

E8A09 (D)
How many levels can an analog-to-digital converter with 8 bit resolution encode?
A. 8
B. 8 multiplied by the gain of the input amplifier
C. 256 divided by the gain of the input amplifier
D. 256

E8A10 (C)
What is the purpose of a low pass filter used in conjunction with a digital-to-analog converter?
A. Lower the input bandwidth to increase the effective resolution
B. Improve accuracy by removing out of sequence codes from the input
C. Remove harmonics from the output caused by the discrete analog levels generated
D. All of these choices are correct

E8A11 (D)
What type of information can be conveyed using digital waveforms?
A. Human speech
B. Video signals
C. Data
D. All of these choices are correct

E8A12 (C)
What is an advantage of using digital signals instead of analog signals to convey the same information?
A. Less complex circuitry is required for digital signal generation and detection
B. Digital signals always occupy a narrower bandwidth
C. Digital signals can be regenerated multiple times without error
D. All of these choices are correct

E8A13 (A)
Which of these methods is commonly used to convert analog signals to digital signals?
A. Sequential sampling
B. Harmonic regeneration
C. Level shifting
D. Phase reversal

E8B - MODULATION AND DEMODULATION: MODULATION METHODS; MODULATION INDEX AND DEVIATION RATIO; FREQUENCY AND TIME DIVISION MULTIPLEXING; ORTHOGONAL FREQUENCY DIVISION MULTIPLEXING

E8B01 (D)
What is the term for the ratio between the frequency deviation of an RF carrier wave and the modulating frequency of its corresponding FM-phone signal?
A. FM compressibility
B. Quieting index
C. Percentage of modulation
D. Modulation index

E8B02 (D)
How does the modulation index of a phase-modulated emission vary with RF carrier frequency (the modulated frequency)?
A. It increases as the RF carrier frequency increases
B. It decreases as the RF carrier frequency increases
C. It varies with the square root of the RF carrier frequency
D. It does not depend on the RF carrier frequency

E8B03 (A)
What is the modulation index of an FM-phone signal having a maximum frequency deviation of 3000 Hz either side of the carrier frequency when the modulating frequency is 1000 Hz?
A. 3
B. 0.3
C. 3000
D. 1000

E8B04 (B)
What is the modulation index of an FM-phone signal having a maximum carrier deviation of plus or minus 6 kHz when modulated with a 2 kHz modulating frequency?
A. 6000
B. 3
C. 2000
D. 1/3

E8B05 (D)
What is the deviation ratio of an FM-phone signal having a maximum frequency swing of plus-or-minus 5 kHz when the maximum modulation frequency is 3 kHz?
A. 60
B. 0.167
C. 0.6
D. 1.67

E8B06 (A)
What is the deviation ratio of an FM-phone signal having a maximum frequency swing of plus or minus 7.5 kHz when the maximum modulation frequency is 3.5 kHz?
A. 2.14
B. 0.214
C. 0.47
D. 47

E8B07 (A)
Orthogonal Frequency Division Multiplexing is a technique used for which type of amateur communication?
A. High speed digital modes
B. Extremely low-power contacts
C. EME
D. OFDM signals are not allowed on amateur bands

E8B08 (D)
What describes Orthogonal Frequency Division Multiplexing?
A. A frequency modulation technique which uses non-harmonically related frequencies
B. A bandwidth compression technique using Fourier transforms
C. A digital mode for narrow band, slow speed transmissions
D. A digital modulation technique using subcarriers at frequencies chosen to avoid intersymbol interference

E8B09 (B)
What is meant by deviation ratio?
A. The ratio of the audio modulating frequency to the center carrier frequency
B. The ratio of the maximum carrier frequency deviation to the highest audio modulating frequency
C. The ratio of the carrier center frequency to the audio modulating frequency
D. The ratio of the highest audio modulating frequency to the average audio modulating frequency

E8B10 (B)
What describes frequency division multiplexing?
A. The transmitted signal jumps from band to band at a predetermined rate
B. Two or more information streams are merged into a baseband, which then modulates the transmitter
C. The transmitted signal is divided into packets of information
D. Two or more information streams are merged into a digital combiner, which then pulse position modulates the transmitter

E8B11 (B)
What is digital time division multiplexing?
A. Two or more data streams are assigned to discrete sub-carriers on an FM transmitter
B. Two or more signals are arranged to share discrete time slots of a data transmission
C. Two or more data streams share the same channel by transmitting time of transmission as the sub-carrier
D. Two or more signals are quadrature modulated to increase bandwidth efficiency

E8C01 (C)
How is Forward Error Correction implemented?
A. By the receiving station repeating each block of three data characters
B. By transmitting a special algorithm to the receiving station along with the data characters
C. By transmitting extra data that may be used to detect and correct transmission errors
D. By varying the frequency shift of the transmitted signal according to a predefined algorithm

E8C02 (C)
What is the definition of symbol rate in a digital transmission?
A. The number of control characters in a message packet
B. The duration of each bit in a message sent over the air
C. The rate at which the waveform of a transmitted signal changes to convey information
D. The number of characters carried per second by the station-to-station link

E8C03 (A)
When performing phase shift keying, why is it advantageous to shift phase precisely at the zero crossing of the RF carrier?
A. This results in the least possible transmitted bandwidth for the particular mode
B. It is easier to demodulate with a conventional, non-synchronous detector
C. It improves carrier suppression
D. All of these choices are correct

E8C04 (C)
What technique is used to minimize the bandwidth requirements of a PSK31 signal?
A. Zero-sum character encoding
B. Reed-Solomon character encoding
C. Use of sinusoidal data pulses
D. Use of trapezoidal data pulses

E8C05 (C)
What is the necessary bandwidth of a 13-WPM international Morse code transmission?
A. Approximately 13 Hz
B. Approximately 26 Hz
C. Approximately 52 Hz
D. Approximately 104 Hz

E8C06 (C)
What is the necessary bandwidth of a 170-hertz shift, 300-baud ASCII transmission?
A. 0.1 Hz
B. 0.3 kHz
C. 0.5 kHz
D. 1.0 kHz

E8C07 (A)
What is the necessary bandwidth of a 4800-Hz frequency shift, 9600-baud ASCII FM transmission?
A. 15.36 kHz
B. 9.6 kHz
C. 4.8 kHz
D. 5.76 kHz

E8C08 (D)
How does ARQ accomplish error correction?
A. Special binary codes provide automatic correction
B. Special polynomial codes provide automatic correction
C. If errors are detected, redundant data is substituted
D. If errors are detected, a retransmission is requested

E8C09 (D)
Which is the name of a digital code where each preceding or following character changes by only one bit?
A. Binary Coded Decimal Code
B. Extended Binary Coded Decimal Interchange Code
C. Excess 3 code
D. Gray code

E8C10 (D)
What is an advantage of Gray code in digital communications where symbols are transmitted as multiple bits
A. It increases security
B. It has more possible states than simple binary
C. It has more resolution than simple binary
D. It facilitates error detection

E8C11 (A)
What is the relationship between symbol rate and baud?
A. They are the same
B. Baud is twice the symbol rate
C. Symbol rate is only used for packet-based modes
D. Baud is only used for RTTY

E8D - KEYING DEFECTS AND OVERMODULATION OF DIGITAL SIGNALS; DIGITAL CODES; SPREAD SPECTRUM

E8D01 (A)
Why are received spread spectrum signals resistant to interference?
A. Signals not using the spread spectrum algorithm are suppressed in the receiver
B. The high power used by a spread spectrum transmitter keeps its signal from being easily overpowered
C. The receiver is always equipped with a digital blanker
D. If interference is detected by the receiver it will signal the transmitter to change frequencies

E8D02 (B)
What spread spectrum communications technique uses a high speed binary bit stream to shift the phase of an RF carrier?
A. Frequency hopping
B. Direct sequence
C. Binary phase-shift keying
D. Phase compandored spread spectrum

E8D03 (D)
How does the spread spectrum technique of frequency hopping work?
A. If interference is detected by the receiver it will signal the transmitter to change frequencies
B. If interference is detected by the receiver it will signal the transmitter to wait until the frequency is clear
C. A pseudo-random binary bit stream is used to shift the phase of an RF carrier very rapidly in a particular sequence
D. The frequency of the transmitted signal is changed very rapidly according to a particular sequence also used by the receiving station

E8D04 (C)
What is the primary effect of extremely short rise or fall time on a CW signal?
A. More difficult to copy
B. The generation of RF harmonics
C. The generation of key clicks
D. Limits data speed

E8D05 (A)
What is the most common method of reducing key clicks?
A. Increase keying waveform rise and fall times
B. Low-pass filters at the transmitter output
C. Reduce keying waveform rise and fall times
D. High-pass filters at the transmitter output

E8D06 (B)
Which of the following indicates likely overmodulation of an AFSK signal such as PSK or MFSK?
A. High reflected power
B. Strong ALC action
C. Harmonics on higher bands
D. Rapid signal fading

E8D07 (D)
What is a common cause of overmodulation of AFSK signals?
A. Excessive numbers of retries
B. Ground loops
C. Bit errors in the modem
D. Excessive transmit audio levels

E8D08 (D)
What parameter might indicate that excessively high input levels are causing distortion in an AFSK signal?
A. Signal to noise ratio
B. Baud rate
C. Repeat Request Rate (RRR)
D. Intermodulation Distortion (IMD)

E8D09 (D)
What is considered a good minimum IMD level for an idling PSK signal?
A. +10 dB
B. +15 dB
C. -20 dB
D. -30 dB

E8D10 (B)
What are some of the differences between the Baudot digital code and ASCII?
A. Baudot uses 4 data bits per character, ASCII uses 7 or 8; Baudot uses 1 character as a letters/figures shift code, ASCII has no letters/figures code
B. Baudot uses 5 data bits per character, ASCII uses 7 or 8; Baudot uses 2 characters as letters/figures shift codes, ASCII has no letters/figures shift code
C. Baudot uses 6 data bits per character, ASCII uses 7 or 8; Baudot has no letters/figures shift code, ASCII uses 2 letters/figures shift codes
D. Baudot uses 7 data bits per character, ASCII uses 8; Baudot has no letters/figures shift code, ASCII uses 2 letters/figures shift codes

E8D11 (C)
What is one advantage of using ASCII code for data communications?
A. It includes built in error correction features
B. It contains fewer information bits per character than any other code
C. It is possible to transmit both upper and lower case text
D. It uses one character as a shift code to send numeric and special characters

E8D12 (D)
What is the advantage of including a parity bit with an ASCII character stream?
A. Faster transmission rate
B. The signal can overpower interfering signals
C. Foreign language characters can be sent
D. Some types of errors can be detected

Subelement E9 - Antennas and Transmission Lines [8 Groups]

E9A - BASIC ANTENNA PARAMETERS: RADIATION RESISTANCE, GAIN, BEAMWIDTH, EFFICIENCY, BEAMWIDTH; EFFECTIVE RADIATED POWER, POLARIZATION

E9A01 (C)
What describes an isotropic antenna?
A. A grounded antenna used to measure earth conductivity
B. A horizontally polarized antenna used to compare Yagi antennas
C. A theoretical antenna used as a reference for antenna gain
D. A spacecraft antenna used to direct signals toward the earth

E9A02 (D)
What antenna has no gain in any direction?
A. Quarter-wave vertical
B. Yagi
C. Half-wave dipole
D. Isotropic antenna

E9A03 (A)
Why would one need to know the feed point impedance of an antenna?
A. To match impedances in order to minimize standing wave ratio on the transmission line
B. To measure the near-field radiation density from a transmitting antenna
C. To calculate the front-to-side ratio of the antenna
D. To calculate the front-to-back ratio of the antenna

E9A04 (B)
Which of the following factors may affect the feed point impedance of an antenna?
A. Transmission-line length
B. Antenna height, conductor length/diameter ratio and location of nearby conductive objects
C. The settings of an antenna tuner at the transmitter
D. Sunspot activity and time of day

E9A05 (D)
What is included in the total resistance of an antenna system?
A. Radiation resistance plus space impedance
B. Radiation resistance plus transmission resistance
C. Transmission-line resistance plus radiation resistance
D. Radiation resistance plus ohmic resistance

E9A06 (D)
How does the beamwidth of an antenna vary as the gain is increased?
A. It increases geometrically
B. It increases arithmetically
C. It is essentially unaffected
D. It decreases

E9A07 (A)
What is meant by antenna gain?
A. The ratio of the radiated signal strength of an antenna in the direction of maximum radiation to that of a reference antenna
B. The ratio of the signal in the forward direction to that in the opposite direction
C. The ratio of the amount of power radiated by an antenna compared to the transmitter output power
D. The final amplifier gain minus the transmission line losses

E9A08 (B)
What is meant by antenna bandwidth?
A. Antenna length divided by the number of elements
B. The frequency range over which an antenna satisfies a performance requirement
C. The angle between the half-power radiation points
D. The angle formed between two imaginary lines drawn through the element ends

E9A09 (B)
How is antenna efficiency calculated?
A. (radiation resistance / transmission resistance) x 100 per cent
B. (radiation resistance / total resistance) x 100 per cent
C. (total resistance / radiation resistance) x 100 per cent
D. (effective radiated power / transmitter output) x 100 percent

E9A10 (A)
Which of the following choices is a way to improve the efficiency of a ground-mounted quarter-wave vertical antenna?
A. Install a good radial system
B. Isolate the coax shield from ground
C. Shorten the radiating element
D. Reduce the diameter of the radiating element

E9A11 (C)
Which of the following factors determines ground losses for a ground-mounted vertical antenna operating in the 3 MHz to 30 MHz range?
A. The standing wave ratio
B. Distance from the transmitter
C. Soil conductivity
D. Take-off angle

E9A12 (A)
How much gain does an antenna have compared to a 1/2-wavelength dipole when it has 6 dB gain over an isotropic antenna?
A. 3.85 dB
B. 6.0 dB
C. 8.15 dB
D. 2.79 dB

E9A13 (B)
How much gain does an antenna have compared to a 1/2-wavelength dipole when it has 12 dB gain over an isotropic antenna?
A. 6.17 dB
B. 9.85 dB
C. 12.5 dB
D. 14.15 dB

E9A14 (C)
What is meant by the radiation resistance of an antenna?
A. The combined losses of the antenna elements and feed line
B. The specific impedance of the antenna
C. The value of a resistance that would dissipate the same amount of power as that radiated from an antenna
D. The resistance in the atmosphere that an antenna must overcome to be able to radiate a signal

E9A15 (D)
What is the effective radiated power relative to a dipole of a repeater station with 150 watts transmitter power output, 2 dB feed line loss, 2.2 dB duplexer loss, and 7 dBd antenna gain?
A. 1977 watts
B. 78.7 watts
C. 420 watts
D. 286 watts

E9A16 (A)
What is the effective radiated power relative to a dipole of a repeater station with 200 watts transmitter power output, 4 dB feed line loss, 3.2 dB duplexer loss, 0.8 dB circulator loss, and 10 dBd antenna gain?
A. 317 watts
B. 2000 watts
C. 126 watts
D. 300 watts

E9A17 (B)
What is the effective radiated power of a repeater station with 200 watts transmitter power output, 2 dB feed line loss, 2.8 dB duplexer loss, 1.2 dB circulator loss, and 7 dBi antenna gain?
A. 159 watts
B. 252 watts
C. 632 watts
D. 63.2 watts

E9A18 (C)
What term describes station output, taking into account all gains and losses?
A. Power factor
B. Half-power bandwidth
C. Effective radiated power
D. Apparent power

E9B - ANTENNA PATTERNS: E AND H PLANE PATTERNS; GAIN AS A FUNCTION OF PATTERN; ANTENNA DESIGN

E9B01 (B)
In the antenna radiation pattern shown in Figure E9-1, what is the 3 dB beam-width?
A. 75 degrees
B. 50 degrees
C. 25 degrees
D. 30 degrees

E9B02 (B)
In the antenna radiation pattern shown in Figure E9-1, what is the front-to-back ratio?
A. 36 dB
B. 18 dB
C. 24 dB
D. 14 dB

E9B03 (B)
In the antenna radiation pattern shown in Figure E9-1, what is the front-to-side ratio?
A. 12 dB
B. 14 dB
C. 18 dB
D. 24 dB

E9B04 (D)
What may occur when a directional antenna is operated at different frequencies within the band for which it was designed?
A. Feed point impedance may become negative
B. The E-field and H-field patterns may reverse
C. Element spacing limits could be exceeded
D. The gain may change depending on frequency

E9B05 (A)
What type of antenna pattern over real ground is shown in Figure E9-2?
A. Elevation
B. Azimuth
C. Radiation resistance
D. Polarization

E9B06 (C)
What is the elevation angle of peak response in the antenna radiation pattern shown in Figure E9-2?
A. 45 degrees
B. 75 degrees
C. 7.5 degrees
D. 25 degrees

E9B07 (C)
How does the total amount of radiation emitted by a directional gain antenna compare with the total amount of radiation emitted from an isotropic antenna, assuming each is driven by the same amount of power?
A. The total amount of radiation from the directional antenna is increased by the gain of the antenna
B. The total amount of radiation from the directional antenna is stronger by its front-to-back ratio
C. They are the same
D. The radiation from the isotropic antenna is 2.15 dB stronger than that from the directional antenna

E9B08 (A)
How can the approximate beam-width in a given plane of a directional antenna be determined?
A. Note the two points where the signal strength of the antenna is 3 dB less than maximum and compute the angular difference
B. Measure the ratio of the signal strengths of the radiated power lobes from the front and rear of the antenna
C. Draw two imaginary lines through the ends of the elements and measure the angle between the lines
D. Measure the ratio of the signal strengths of the radiated power lobes from the front and side of the antenna

E9B09 (B)
What type of computer program technique is commonly used for modeling antennas?
A. Graphical analysis
B. Method of Moments
C. Mutual impedance analysis
D. Calculus differentiation with respect to physical properties

E9B10 (A)
What is the principle of a Method of Moments analysis?
A. A wire is modeled as a series of segments, each having a uniform value of current
B. A wire is modeled as a single sine-wave current generator
C. A wire is modeled as a series of points, each having a distinct location in space
D. A wire is modeled as a series of segments, each having a distinct value of voltage across it

E9B11 (C)
What is a disadvantage of decreasing the number of wire segments in an antenna model below the guideline of 10 segments per half-wavelength?
A. Ground conductivity will not be accurately modeled
B. The resulting design will favor radiation of harmonic energy
C. The computed feed point impedance may be incorrect
D. The antenna will become mechanically unstable

E9B12 (D)
What is the far field of an antenna?
A. The region of the ionosphere where radiated power is not refracted
B. The region where radiated power dissipates over a specified time period
C. The region where radiated field strengths are obstructed by objects of reflection
D. The region where the shape of the antenna pattern is independent of distance

E9B13 (B)
What does the abbreviation NEC stand for when applied to antenna modeling programs?
A. Next Element Comparison
B. Numerical Electromagnetic Code
C. National Electrical Code
D. Numeric Electrical Computation

E9B14 (D)
What type of information can be obtained by submitting the details of a proposed new antenna to a modeling program?
A. SWR vs frequency charts
B. Polar plots of the far field elevation and azimuth patterns
C. Antenna gain
D. All of these choices are correct

E9B15 (B)
What is the front-to-back ratio of the radiation pattern shown in Figure E9-2?
A. 15 dB
B. 28 dB
C. 3 dB
D. 24 dB

E9B16 (A)
How many elevation lobes appear in the forward direction of the antenna radiation pattern shown in Figure E9-2?
A. 4
B. 3
C. 1
D. 7

E9C - WIRE AND PHASED ARRAY ANTENNAS: RHOMBIC ANTENNAS; EFFECTS OF GROUND REFLECTIONS; E-OFF ANGLES; PRACTICAL WIRE ANTENNAS: ZEPPS, OCFD, LOOPS

E9C01 (D)
What is the radiation pattern of two 1/4-wavelength vertical antennas spaced 1/2-wavelength apart and fed 180 degrees out of phase?
A. Cardioid
B. Omni-directional
C. A figure-8 broadside to the axis of the array
D. A figure-8 oriented along the axis of the array

E9C02 (A)
What is the radiation pattern of two 1/4 wavelength vertical antennas spaced 1/4 wavelength apart and fed 90 degrees out of phase?
A. Cardioid
B. A figure-8 end-fire along the axis of the array
C. A figure-8 broadside to the axis of the array
D. Omni-directional

E9C03 (C)
What is the radiation pattern of two 1/4 wavelength vertical antennas spaced a 1/2 wavelength apart and fed in phase?
A. Omni-directional
B. Cardioid
C. A Figure-8 broadside to the axis of the array
D. A Figure-8 end-fire along the axis of the array

E9C04 (B)
What happens to the radiation pattern of an unterminated long wire antenna as the wire length is increased?
A. The lobes become more perpendicular to the wire
B. The lobes align more in the direction of the wire
C. The vertical angle increases
D. The front-to-back ratio decreases

E9C05 (A)
What is an OCFD antenna?
A. A dipole feed approximately 1/3 the way from one end with a 4:1 balun to provide multiband operation
B. A remotely tunable dipole antenna using orthogonally controlled frequency diversity
C. An eight band dipole antenna using octophase filters
D. A multiband dipole antenna using one-way circular polarization for frequency diversity

E9C06 (B)
What is the effect of a terminating resistor on a rhombic antenna?
A. It reflects the standing waves on the antenna elements back to the transmitter
B. It changes the radiation pattern from bidirectional to unidirectional
C. It changes the radiation pattern from horizontal to vertical polarization
D. It decreases the ground loss

E9C07 (A)
What is the approximate feed point impedance at the center of a two-wire folded dipole antenna?
A. 300 ohms
B. 72 ohms
C. 50 ohms
D. 450 ohms

E9C08 (C)
What is a folded dipole antenna?
A. A dipole one-quarter wavelength long
B. A type of ground-plane antenna
C. A dipole consisting of one wavelength of wire forming a very thin loop
D. A dipole configured to provide forward gain

E9C09 (A)
What is a G5RV antenna?
A. A multi-band dipole antenna fed with coax and a balun through a selected length of open wire transmission line
B. A multi-band trap antenna
C. A phased array antenna consisting of multiple loops
D. A wide band dipole using shorted coaxial cable for the radiating elements and fed with a 4:1 balun

E9C10 (B)
Which of the following describes a Zepp antenna?
A. A dipole constructed from zip cord
B. An end fed dipole antenna
C. An omni-directional antenna commonly used for satellite communications
D. A vertical array capable of quickly changing the direction of maximum radiation by changing phasing lines

E9C11 (D)
How is the far-field elevation pattern of a vertically polarized antenna affected by being mounted over seawater versus rocky ground?
A. The low-angle radiation decreases
B. The high-angle radiation increases
C. Both the high-angle and low-angle radiation decrease
D. The low-angle radiation increases

E9C12 (C)
Which of the following describes an extended double Zepp antenna?
A. A wideband vertical antenna constructed from precisely tapered aluminum tubing
B. A portable antenna erected using two push support poles
C. A center fed 1.25 wavelength antenna (two 5/8 wave elements in phase)
D. An end fed folded dipole antenna

E9C13 (C)
What is the main effect of placing a vertical antenna over an imperfect ground?
A. It causes increased SWR
B. It changes the impedance angle of the matching network
C. It reduces low-angle radiation
D. It reduces losses in the radiating portion of the antenna

E9C14 (B)
How does the performance of a horizontally polarized antenna mounted on the side of a hill compare with the same antenna mounted on flat ground?
A. The main lobe takeoff angle increases in the downhill direction
B. The main lobe takeoff angle decreases in the downhill direction
C. The horizontal beam width decreases in the downhill direction
D. The horizontal beam width increases in the uphill direction

E9C15 (B)
How does the radiation pattern of a horizontally polarized 3-element beam antenna vary with its height above ground?
A. The main lobe takeoff angle increases with increasing height
B. The main lobe takeoff angle decreases with increasing height
C. The horizontal beam width increases with height
D. The horizontal beam width decreases with height

E9D01 (C)
How does the gain of an ideal parabolic dish antenna change when the operating frequency is doubled?
A. Gain does not change
B. Gain is multiplied by 0.707
C. Gain increases by 6 dB
D. Gain increases by 3 dB

E9D02 (C)
How can linearly polarized Yagi antennas be used to produce circular polarization?
A. Stack two Yagis fed 90 degrees out of phase to form an array with the respective elements in parallel planes
B. Stack two Yagis fed in phase to form an array with the respective elements in parallel planes
C. Arrange two Yagis perpendicular to each other with the driven elements at the same point on the boom fed 90 degrees out of phase
D. Arrange two Yagis collinear to each other with the driven elements fed 180 degrees out of phase

E9D03 (A)
Where should a high Q loading coil be placed to minimize losses in a shortened vertical antenna?
A. Near the center of the vertical radiator
B. As low as possible on the vertical radiator
C. As close to the transmitter as possible
D. At a voltage node

E9D04 (C)
Why should an HF mobile antenna loading coil have a high ratio of reactance to resistance?
A. To swamp out harmonics
B. To maximize losses
C. To minimize losses
D. To minimize the Q

E9D05 (A)
What is a disadvantage of using a multiband trapped antenna?
A. It might radiate harmonics
B. It radiates the harmonics and fundamental equally well
C. It is too sharply directional at lower frequencies
D. It must be neutralized

E9D06 (B)
What happens to the bandwidth of an antenna as it is shortened through the use of loading coils?
A. It is increased
B. It is decreased
C. No change occurs
D. It becomes flat

E9D07 (D)
What is an advantage of using top loading in a shortened HF vertical antenna?
A. Lower Q
B. Greater structural strength
C. Higher losses
D. Improved radiation efficiency

E9D08 (B)
What happens as the Q of an antenna increases?
A. SWR bandwidth increases
B. SWR bandwidth decreases
C. Gain is reduced
D. More common-mode current is present on the feed line

E9D09 (D)
What is the function of a loading coil used as part of an HF mobile antenna?
A. To increase the SWR bandwidth
B. To lower the losses
C. To lower the Q
D. To cancel capacitive reactance

E9D10 (B)
What happens to feed point impedance at the base of a fixed length HF mobile antenna as the frequency of operation is lowered?
A. The radiation resistance decreases and the capacitive reactance decreases
B. The radiation resistance decreases and the capacitive reactance increases
C. The radiation resistance increases and the capacitive reactance decreases
D. The radiation resistance increases and the capacitive reactance increases

E9D11 (B)
Which of the following types of conductors would be best for minimizing losses in a station's RF ground system?
A. A resistive wire, such as spark plug wire
B. A wide flat copper strap
C. A cable with six or seven 18 gauge conductors in parallel
D. A single 12 gauge or 10 gauge stainless steel wire

E9D12 (C)
Which of the following would provide the best RF ground for your station?
A. A 50 ohm resistor connected to ground
B. An electrically short connection to a metal water pipe
C. An electrically short connection to 3 or 4 interconnected ground rods driven into the Earth
D. An electrically short connection to 3 or 4 interconnected ground rods via a series RF choke

E9D13 (B)
What usually occurs if a Yagi antenna is designed solely for maximum forward gain?
A. The front-to-back ratio increases
B. The front-to-back ratio decreases
C. The frequency response is widened over the whole frequency band
D. The SWR is reduced

E9E - MATCHING: MATCHING ANTENNAS TO FEED LINES; PHASING LINES; POWER DIVIDERS

E9E01 (B)
What system matches a higher impedance transmission line to a lower impedance antenna by connecting the line to the driven element in two places spaced a fraction of a wavelength each side of element center?
A. The gamma matching system
B. The delta matching system
C. The omega matching system
D. The stub matching system

E9E02 (A)
What is the name of an antenna matching system that matches an unbalanced feed line to an antenna by feeding the driven element both at the center of the element and at a fraction of a wavelength to one side of center?
A. The gamma match
B. The delta match
C. The epsilon match
D. The stub match

E9E03 (D)
What is the name of the matching system that uses a section of transmission line connected in parallel with the feed line at or near the feed point?
A. The gamma match
B. The delta match
C. The omega match
D. The stub match

E9E04 (B)
What is the purpose of the series capacitor in a gamma-type antenna matching network?
A. To provide DC isolation between the feed line and the antenna
B. To cancel the inductive reactance of the matching network
C. To provide a rejection notch that prevents the radiation of harmonics
D. To transform the antenna impedance to a higher value

E9E05 (A)
How must the driven element in a 3-element Yagi be tuned to use a hairpin matching system?
A. The driven element reactance must be capacitive
B. The driven element reactance must be inductive
C. The driven element resonance must be lower than the operating frequency
D. The driven element radiation resistance must be higher than the characteristic impedance of the transmission line

E9E06 (C)
What is the equivalent lumped-constant network for a hairpin matching system of a 3-element Yagi?
A. Pi-network
B. Pi-L-network
C. A shunt inductor
D. A series capacitor

E9E07 (B)
What term best describes the interactions at the load end of a mismatched transmission line?
A. Characteristic impedance
B. Reflection coefficient
C. Velocity factor
D. Dielectric constant

E9E08 (D)
Which of the following measurements is characteristic of a mismatched transmission line?
A. An SWR less than 1:1
B. A reflection coefficient greater than 1
C. A dielectric constant greater than 1
D. An SWR greater than 1:1

E9E09 (C)
Which of these matching systems is an effective method of connecting a 50 ohm coaxial cable feed line to a grounded tower so it can be used as a vertical antenna?
A. Double-bazooka match
B. Hairpin match
C. Gamma match
D. All of these choices are correct

E9E10 (C)
Which of these choices is an effective way to match an antenna with a 100 ohm feed point impedance to a 50 ohm coaxial cable feed line?
A. Connect a 1/4-wavelength open stub of 300 ohm twin-lead in parallel with the coaxial feed line where it connects to the antenna
B. Insert a 1/2 wavelength piece of 300 ohm twin-lead in series between the antenna terminals and the 50 ohm feed cable
C. Insert a 1/4-wavelength piece of 75 ohm coaxial cable transmission line in series between the antenna terminals and the 50 ohm feed cable
D. Connect 1/2 wavelength shorted stub of 75 ohm cable in parallel with the 50 ohm cable where it attaches to the antenna

E9E11 (B)
What is an effective way of matching a feed line to a VHF or UHF antenna when the impedances of both the antenna and feed line are unknown?
A. Use a 50 ohm 1:1 balun between the antenna and feed line
B. Use the universal stub matching technique
C. Connect a series-resonant LC network across the antenna feed terminals
D. Connect a parallel-resonant LC network across the antenna feed terminals

E9E12 (A)
What is the primary purpose of a phasing line when used with an antenna having multiple driven elements?
A. It ensures that each driven element operates in concert with the others to create the desired antenna pattern
B. It prevents reflected power from traveling back down the feed line and causing harmonic radiation from the transmitter
C. It allows single-band antennas to operate on other bands
D. It makes sure the antenna has a low-angle radiation pattern

E9E13 (C)
What is a use for a Wilkinson divider?
A. It divides the operating frequency of a transmitter signal so it can be used on a lower frequency band
B. It is used to feed high-impedance antennas from a low-impedance source
C. It is used to divide power equally between two 50 ohm loads while maintaining 50 ohm input impedance
D. It is used to feed low-impedance loads from a high-impedance source

E9F - TRANSMISSION LINES: CHARACTERISTICS OF OPEN AND SHORTED FEED LINES; 1/8 WAVELENGTH; 1/4 WAVELENGTH; 1/2 WAVELENGTH; FEED LINES: COAX VERSUS OPEN-WIRE; VELOCITY FACTOR; ELECTRICAL LENGTH; COAXIAL CABLE DIELECTRICS; VELOCITY FACTOR

E9F01 (D)
What is the velocity factor of a transmission line?
A. The ratio of the characteristic impedance of the line to the terminating impedance
B. The index of shielding for coaxial cable
C. The velocity of the wave in the transmission line multiplied by the velocity of light in a vacuum
D. The velocity of the wave in the transmission line divided by the velocity of light in a vacuum

E9F02 (C)
Which of the following determines the velocity factor of a transmission line?
A. The termination impedance
B. The line length
C. Dielectric materials used in the line
D. The center conductor resistivity

E9F03 (D)
Why is the physical length of a coaxial cable transmission line shorter than its electrical length?
A. Skin effect is less pronounced in the coaxial cable
B. The characteristic impedance is higher in a parallel feed line
C. The surge impedance is higher in a parallel feed line
D. Electrical signals move more slowly in a coaxial cable than in air

E9F04 (B)
What is the typical velocity factor for a coaxial cable with solid polyethylene dielectric?
A. 2.70
B. 0.66
C. 0.30
D. 0.10

E9F05 (C)
What is the approximate physical length of a solid polyethylene dielectric coaxial transmission line that is electrically one-quarter wavelength long at 14.1 MHz?
A. 20 meters
B. 2.3 meters
C. 3.5 meters
D. 0.2 meters

E9F06 (C)
What is the approximate physical length of an air-insulated, parallel conductor transmission line that is electrically one-half wavelength long at 14.10 MHz?
A. 15 meters
B. 20 meters
C. 10 meters
D. 71 meters

E9F07 (A)
How does ladder line compare to small-diameter coaxial cable such as RG-58 at 50 MHz?
A. Lower loss
B. Higher SWR
C. Smaller reflection coefficient
D. Lower velocity factor

E9F08 (A)
What is the term for the ratio of the actual speed at which a signal travels through a transmission line to the speed of light in a vacuum?
A. Velocity factor
B. Characteristic impedance
C. Surge impedance
D. Standing wave ratio

E9F09 (B)
What is the approximate physical length of a solid polyethylene dielectric coaxial transmission line that is electrically one-quarter wavelength long at 7.2 MHz?
A. 10 meters
B. 6.9 meters
C. 24 meters
D. 50 meters

E9F10 (C)
What impedance does a 1/8 wavelength transmission line present to a generator when the line is shorted at the far end?
A. A capacitive reactance
B. The same as the characteristic impedance of the line
C. An inductive reactance
D. The same as the input impedance to the final generator stage

E9F11 (C)
What impedance does a 1/8 wavelength transmission line present to a generator when the line is open at the far end?
A. The same as the characteristic impedance of the line
B. An inductive reactance
C. A capacitive reactance
D. The same as the input impedance of the final generator stage

E9F12 (D)
What impedance does a 1/4 wavelength transmission line present to a generator when the line is open at the far end?
A. The same as the characteristic impedance of the line
B. The same as the input impedance to the generator
C. Very high impedance
D. Very low impedance

E9F13 (A)
What impedance does a 1/4 wavelength transmission line present to a generator when the line is shorted at the far end?
A. Very high impedance
B. Very low impedance
C. The same as the characteristic impedance of the transmission line
D. The same as the generator output impedance

E9F14 (B)
What impedance does a 1/2 wavelength transmission line present to a generator when the line is shorted at the far end?
A. Very high impedance
B. Very low impedance
C. The same as the characteristic impedance of the line
D. The same as the output impedance of the generator

E9F15 (A)
What impedance does a 1/2 wavelength transmission line present to a generator when the line is open at the far end?
A. Very high impedance
B. Very low impedance
C. The same as the characteristic impedance of the line
D. The same as the output impedance of the generator

E9F16 (D)
Which of the following is a significant difference between foam dielectric coaxial cable and solid dielectric cable, assuming all other parameters are the same?
A. Foam dielectric has lower safe operating voltage limits
B. Foam dielectric has lower loss per unit of length
C. Foam dielectric has higher velocity factor
D. All of these choices are correct

E9G - THE SMITH CHART

E9G01 (A)
Which of the following can be calculated using a Smith chart?
A. Impedance along transmission lines
B. Radiation resistance
C. Antenna radiation pattern
D. Radio propagation

E9G02 (B)
What type of coordinate system is used in a Smith chart?
A. Voltage circles and current arcs
B. Resistance circles and reactance arcs
C. Voltage lines and current chords
D. Resistance lines and reactance chords

E9G03 (C)
Which of the following is often determined using a Smith chart?
A. Beam headings and radiation patterns
B. Satellite azimuth and elevation bearings
C. Impedance and SWR values in transmission lines
D. Trigonometric functions

E9G04 (C)
What are the two families of circles and arcs that make up a Smith chart?
A. Resistance and voltage
B. Reactance and voltage
C. Resistance and reactance
D. Voltage and impedance

E9G05 (A)
What type of chart is shown in Figure E9-3?
A. Smith chart
B. Free space radiation directivity chart
C. Elevation angle radiation pattern chart
D. Azimuth angle radiation pattern chart

E9G06 (B)
On the Smith chart shown in Figure E9-3, what is the name for the large outer circle on which the reactance arcs terminate?
A. Prime axis
B. Reactance axis
C. Impedance axis
D. Polar axis

E9G07 (D)
On the Smith chart shown in Figure E9-3, what is the only straight line shown?
A. The reactance axis
B. The current axis
C. The voltage axis
D. The resistance axis

E9G08 (C)
What is the process of normalization with regard to a Smith chart?
A. Reassigning resistance values with regard to the reactance axis
B. Reassigning reactance values with regard to the resistance axis
C. Reassigning impedance values with regard to the prime center
D. Reassigning prime center with regard to the reactance axis

E9G09 (A)
What third family of circles is often added to a Smith chart during the process of solving problems?
A. Standing wave ratio circles
B. Antenna-length circles
C. Coaxial-length circles
D. Radiation-pattern circles

E9G10 (D)
What do the arcs on a Smith chart represent?
A. Frequency
B. SWR
C. Points with constant resistance
D. Points with constant reactance

E9G11 (B)
How are the wavelength scales on a Smith chart calibrated?
A. In fractions of transmission line electrical frequency
B. In fractions of transmission line electrical wavelength
C. In fractions of antenna electrical wavelength
D. In fractions of antenna electrical frequency

E9H - RECEIVING ANTENNAS: RADIO DIRECTION FINDING ANTENNAS; BEVERAGE ANTENNAS; SPECIALIZED RECEIVING ANTENNAS; LONGWIRE RECEIVING ANTENNAS

E9H01 (D)
When constructing a Beverage antenna, which of the following factors should be included in the design to achieve good performance at the desired frequency?
A. Its overall length must not exceed 1/4 wavelength
B. It must be mounted more than 1 wavelength above ground
C. It should be configured as a four-sided loop
D. It should be one or more wavelengths long

E9H02 (A)
Which is generally true for low band (160 meter and 80 meter) receiving antennas?
A. Atmospheric noise is so high that gain over a dipole is not important
B. They must be erected at least 1/2 wavelength above the ground to attain good directivity
C. Low loss coax transmission line is essential for good performance
D. All of these choices are correct

E9H03 DELETED February 1, 2016

E9H04 (B)
What is an advantage of using a shielded loop antenna for direction finding?
A. It automatically cancels ignition noise in mobile installations
B. It is electro statically balanced against ground, giving better nulls
C. It eliminates tracking errors caused by strong out-of-band signals
D. It allows stations to communicate without giving away their position

E9H05 (A)
What is the main drawback of a wire-loop antenna for direction finding?
A. It has a bidirectional pattern
B. It is non-rotatable
C. It receives equally well in all directions
D. It is practical for use only on VHF bands

E9H06 (C)
What is the triangulation method of direction finding?
A. The geometric angles of sky waves from the source are used to determine its position
B. A fixed receiving station plots three headings to the signal source
C. Antenna headings from several different receiving locations are used to locate the signal source
D. A fixed receiving station uses three different antennas to plot the location of the signal source

E9H07 (D)
Why is it advisable to use an RF attenuator on a receiver being used for direction finding?
A. It narrows the bandwidth of the received signal to improve signal to noise ratio
B. It compensates for the effects of an isotropic antenna, thereby improving directivity
C. It reduces loss of received signals caused by antenna pattern nulls, thereby increasing sensitivity
D. It prevents receiver overload which could make it difficult to determine peaks or nulls

E9H08 (A)
What is the function of a sense antenna?
A. It modifies the pattern of a DF antenna array to provide a null in one direction
B. It increases the sensitivity of a DF antenna array
C. It allows DF antennas to receive signals at different vertical angles
D. It provides diversity reception that cancels multipath signals

E9H09 (C)
Which of the following describes the construction of a receiving loop antenna?
A. A large circularly polarized antenna
B. A small coil of wire tightly wound around a toroidal ferrite core
C. One or more turns of wire wound in the shape of a large open coil
D. A vertical antenna coupled to a feed line through an inductive loop of wire

E9H10 (D)
How can the output voltage of a multiple turn receiving loop antenna be increased?
A. By reducing the permeability of the loop shield
B. By increasing the number of wire turns in the loop and reducing the area of the loop structure
C. By winding adjacent turns in opposing directions
D. By increasing either the number of wire turns in the loop or the area of the loop structure or both

E9H11 (B)
What characteristic of a cardioid pattern antenna is useful for direction finding?
A. A very sharp peak
B. A very sharp single null
C. Broad band response
D. High-radiation angle

Subelement E0 – Safety - [1 Group]

E0A - SAFETY: AMATEUR RADIO SAFETY PRACTICES; RF RADIATION HAZARDS; HAZARDOUS MATERIALS; GROUNDING

E0A01 (B)
What is the primary function of an external earth connection or ground rod?
A. Reduce received noise
B. Lightning protection
C. Reduce RF current flow between pieces of equipment
D. Reduce RFI to telephones and home entertainment systems

E0A02 (B)
When evaluating RF exposure levels from your station at a neighbor's home, what must you do?
A. Make sure signals from your station are less than the controlled MPE limits
B. Make sure signals from your station are less than the uncontrolled MPE limits
C. You need only evaluate exposure levels on your own property
D. Advise your neighbors of the results of your tests

E0A03 (C)
Which of the following would be a practical way to estimate whether the RF fields produced by an amateur radio station are within permissible MPE limits?
A. Use a calibrated antenna analyzer
B. Use a hand calculator plus Smith-chart equations to calculate the fields
C. Use an antenna modeling program to calculate field strength at accessible locations
D. All of the choices are correct

E0A04 (C)
When evaluating a site with multiple transmitters operating at the same time, the operators and licensees of which transmitters are responsible for mitigating over-exposure situations?
A. Only the most powerful transmitter
B. Only commercial transmitters
C. Each transmitter that produces 5 percent or more of its MPE limit at accessible locations
D. Each transmitter operating with a duty-cycle greater than 50 percent

E0A05 (B)
What is one of the potential hazards of using microwaves in the amateur radio bands?
A. Microwaves are ionizing radiation
B. The high gain antennas commonly used can result in high exposure levels
C. Microwaves often travel long distances by ionospheric reflection
D. The extremely high frequency energy can damage the joints of antenna structures

E0A06 (D)
Why are there separate electric (E) and magnetic (H) field MPE limits?
A. The body reacts to electromagnetic radiation from both the E and H fields
B. Ground reflections and scattering make the field impedance vary with location
C. E field and H field radiation intensity peaks can occur at different locations
D. All of these choices are correct

E0A07 (B)
How may dangerous levels of carbon monoxide from an emergency generator be detected?
A. By the odor
B. Only with a carbon monoxide detector
C. Any ordinary smoke detector can be used
D. By the yellowish appearance of the gas

E0A08 (C)
What does SAR measure?
A. Synthetic Aperture Ratio of the human body
B. Signal Amplification Rating
C. The rate at which RF energy is absorbed by the body
D. The rate of RF energy reflected from stationary terrain

E0A09 (C)
Which insulating material commonly used as a thermal conductor for some types of electronic devices is extremely toxic if broken or crushed and the particles are accidentally inhaled?
A. Mica
B. Zinc oxide
C. Beryllium Oxide
D. Uranium Hexafluoride

E0A10 (A)
What toxic material may be present in some electronic components such as high voltage capacitors and transformers?
A. Polychlorinated Biphenyls
B. Polyethylene
C. Polytetrafluroethylene
D. Polymorphic silicon

E0A11 (C)
Which of the following injuries can result from using high-power UHF or microwave transmitters?
A. Hearing loss caused by high voltage corona discharge
B. Blood clotting from the intense magnetic field
C. Localized heating of the body from RF exposure in excess of the MPE limits
D. Ingestion of ozone gas from the cooling system

~~~~End of question pool text~~~~

Final Review

Your main study effort now behind you, you can relax—but not totally. Now is the time for some (albeit relaxed) final review using the "Quick Study" Streamlined Question Pools. These are extremely valuable study techniques to make your review process faster and easier, especially the last days before the exam—or the morning of exam day, while gulping down a last-minute cup of coffee. You have three different Streamlined Question Pools to choose from:

1. **Questions with Just the Right Answers.** As the test date draws nearer, reviewing the questions with "Just the Right Answers" and not all the wrong choices allows a comprehensive yet *quicker* review of the test material.
2. **Just the Right Answers without the Questions.** A quick scan of only the right answers, without the questions is surprisingly helpful to unstick a stalled brain in the middle of the exam. One Extra Class Amateur Radio Operator emailed me that when one of the exam questions stumped him, he remembered having seen it on "Just the Right Answer" list and it helped him pass!
3. **Just the Questions without the Answers.** Finally, to test your learning, "Just the Questions" work like flashcards to help you test your knowledge without benefit of seeing any of the multiple choice questions.

Good luck! You are almost there!

QUESTIONS WITH JUST THE RIGHT ANSWERS

Below are all the multiple choice questions in the Extra Class Question Pool but edited to include only the right answers and omitting all the wrong ones. The answers are bolded for even quicker scanning.

Subelement E1 – Commission's Rules [6 Groups]

E1A - OPERATING STANDARDS: FREQUENCY PRIVILEGES; EMISSION STANDARDS; AUTOMATIC MESSAGE FORWARDING; FREQUENCY SHARING; STATIONS ABOARD SHIPS OR AIRCRAFT

When using a transceiver that displays the carrier frequency of phone signals, which of the following displayed frequencies represents the highest frequency at which a properly adjusted USB emission will be totally within the band?
✓ **3 kHz below the upper band edge**

When using a transceiver that displays the carrier frequency of phone signals, which of the following displayed frequencies represents the lowest frequency at which a properly adjusted LSB emission will be totally within the band?
✓ **3 kHz above the lower band edge**

With your transceiver displaying the carrier frequency of phone signals, you hear a station calling CQ on 14.349 MHz USB. Is it legal to return the call using upper sideband on the same frequency?
✓ **No, the sideband will extend beyond the band edge**

With your transceiver displaying the carrier frequency of phone signals, you hear a DX station calling CQ on 3.601 MHz LSB. Is it legal to return the call using lower sideband on the same frequency?
✓ **No, the sideband will extend beyond the edge of the phone band segment**

What is the maximum power output permitted on the 60 meter band?
✓ **100 watts PEP effective radiated power relative to the gain of a half-wave dipole**

Where must the carrier frequency of a CW signal be set to comply with FCC rules for 60 meter operation?
✓ **At the center frequency of the channel**

Which amateur band requires transmission on specific channels rather than on a range of frequencies?
✓ **60 meter band**

If a station in a message forwarding system inadvertently forwards a message that is in violation of FCC rules, who is primarily accountable for the rules violation?
✓ **The control operator of the originating station**

What is the first action you should take if your digital message forwarding station inadvertently forwards a communication that violates FCC rules?
✓ **Discontinue forwarding the communication as soon as you become aware of it**

If an amateur station is installed aboard a ship or aircraft, what condition must be met before the station is operated?
✓ **Its operation must be approved by the master of the ship or the pilot in command of the aircraft**

Which of the following describes authorization or licensing required when operating an amateur station aboard a U.S.-registered vessel in international waters?
✓ **Any FCC-issued amateur license**

With your transceiver displaying the carrier frequency of CW signals, you hear a DX station's CQ on 3.500 MHz. Is it legal to return the call using CW on the same frequency?
✓ **No, one of the sidebands of the CW signal will be out of the band**

Who must be in physical control of the station apparatus of an amateur station aboard any vessel or craft that is documented or registered in the United States?
✓ **Any person holding an FCC issued amateur license or who is authorized for alien reciprocal operation**

What is the maximum bandwidth for a data emission on 60 meters?
✓ **2.8 kHz**

E1B – STATION RESTRICTIONS AND SPECIAL OPERATIONS: RESTRICTIONS ON STATION LOCATION; GENERAL OPERATING RESTRICTIONS, SPURIOUS EMISSIONS, CONTROL OPERATOR REIMBURSEMENT; ANTENNA STRUCTURE RESTRICTIONS; RACES OPERATIONS; NATIONAL QUIET ZONE

Which of the following constitutes a spurious emission?
✓ **An emission outside its necessary bandwidth that can be reduced or eliminated without affecting the information transmitted**

Which of the following factors might cause the physical location of an amateur station apparatus or antenna structure to be restricted?
✓ **The location is of environmental importance or significant in American history, architecture, or culture**

Within what distance must an amateur station protect an FCC monitoring facility from harmful interference?
✓ **1 mile**

What must be done before placing an amateur station within an officially designated wilderness area or wildlife preserve, or an area listed in the National Register of Historical Places?

✓ **An Environmental Assessment must be submitted to the FCC**

What is the National Radio Quiet Zone?

✓ **An area surrounding the National Radio Astronomy Observatory**

Which of the following additional rules apply if you are installing an amateur station antenna at a site at or near a public use airport?

✓ **You may have to notify the Federal Aviation Administration and register it with the FCC as required by Part 17 of FCC rules**

What is the highest modulation index permitted at the highest modulation frequency for angle modulation below 29.0 MHz?

✓ **1.0**

What limitations may the FCC place on an amateur station if its signal causes interference to domestic broadcast reception, assuming that the receivers involved are of good engineering design?

✓ **The amateur station must avoid transmitting during certain hours on frequencies that cause the interference**

Which amateur stations may be operated under RACES rules?

✓ **Any FCC-licensed amateur station certified by the responsible civil defense organization for the area served**

What frequencies are authorized to an amateur station operating under RACES rules?

✓ **All amateur service frequencies authorized to the control operator**

What is the permitted mean power of any spurious emission relative to the mean power of the fundamental emission from a station transmitter or external RF amplifier installed after January 1, 2003 and transmitting on a frequency below 30 MHZ?

✓ **At least 43 dB below**

E1C – DEFINITIONS AND RESTRICTIONS PERTIANING TO LOCAL, AUTOMATIC AND REMOTE CONTROL OPERATION; CONTROL OPERATOR RESPONSIBILITIES FOR REMOTE AND AUTOMATICALLY CONTROLLED STATIONS; IARP AND CEPT LICENSES; THIRD PARTY COMMUNICATIONS OVER AUTOMATICALLY CONTROLLED STATIONS

What is a remotely controlled station?

✓ **A station controlled indirectly through a control link**

What is meant by automatic control of a station?

✓ **The use of devices and procedures for control so that the control operator does not have to be present at a control point**

How do the control operator responsibilities of a station under automatic control differ from one under local control?

✓ **Under automatic control the control operator is not required to be present at the control point**

What is meant by IARP?

✓ **An international amateur radio permit that allows U.S. amateurs to operate in certain countries of the Americas**

When may an automatically controlled station originate third party communications?

✓ **Never**

Which of the following statements concerning remotely controlled amateur stations is true?

✓ **A control operator must be present at the control point**

What is meant by local control?

✓ **Direct manipulation of the transmitter by a control operator**

What is the maximum permissible duration of a remotely controlled station's transmissions if its control link malfunctions?

✓ **3 minutes**

Which of these ranges of frequencies is available for an automatically controlled repeater operating below 30 MHz?

✓ **29.500 MHz - 29.700 MHz**

What types of amateur stations may automatically retransmit the radio signals of other amateur stations?

✓ **Only auxiliary, repeater or space stations**

Which of the following operating arrangements allows an FCC-licensed U.S. citizen to operate in many European countries, and alien amateurs from many European countries to operate in the U.S.?

✓ **CEPT agreement**

What types of communications may be transmitted to amateur stations in foreign countries?

✓ **Communications incidental to the purpose of the amateur service and remarks of a personal nature**

Which of the following is required in order to operate in accordance with CEPT rules in foreign countries where permitted?

✓ **You must bring a copy of FCC Public Notice DA 11-221**

<u>E1D – AMATEUR SATELLITES; DEFINITIONS AND PURPOSE; LICENSE REQURIEMENTS FOR SPACE STATIONS; AVAILABLE FREQUENCIES AND BANDS; TELECOMMAND AND TELEMETRY OPERATIONS; RESTRICTIONS, AND SPECIAL PROVISIONS; NOTIFICATION REQUIREMENTS</u>

What is the definition of the term telemetry?
✓ **One-way transmission of measurements at a distance from the measuring instrument**

What is the amateur satellite service?
✓ **A radio communications service using amateur radio stations on satellites**

What is a telecommand station in the amateur satellite service?
✓ **An amateur station that transmits communications to initiate, modify or terminate functions of a space station**

What is an Earth station in the amateur satellite service?
✓ **An amateur station within 50 km of the Earth's surface intended for communications with amateur stations by means of objects in space**

What class of licensee is authorized to be the control operator of a space station?
✓ **Any class with appropriate operator privileges**

Which of the following is a requirement of a space station?
✓ **The space station must be capable of terminating transmissions by telecommand when directed by the FCC**

Which amateur service HF bands have frequencies authorized for space stations?
✓ **Only the 40 m, 20 m, 17 m, 15 m, 12 m and 10 m bands**

Which VHF amateur service bands have frequencies available for space stations?
✓ **2 meters**

Which UHF amateur service bands have frequencies available for a space station?
✓ **70 cm and 13 cm**

Which amateur stations are eligible to be telecommand stations?
✓ **Any amateur station so designated by the space station licensee, subject to the privileges of the class of operator license held by the control operator**

Which amateur stations are eligible to operate as Earth stations?
✓ **Any amateur station, subject to the privileges of the class of operator license held by the control operator**

What is the minimum number of qualified VEs required to administer an Element 4 amateur operator license examination?
- ✓ **3**

Where are the questions for all written U.S. amateur license examinations listed?
- ✓ **In a question pool maintained by all the VECs**

What is a Volunteer Examiner Coordinator?
- ✓ **An organization that has entered into an agreement with the FCC to coordinate amateur operator license examinations**

Which of the following best describes the Volunteer Examiner accreditation process?
- ✓ **The procedure by which a VEC confirms that the VE applicant meets FCC requirements to serve as an examiner**

What is the minimum passing score on amateur operator license examinations?
- ✓ **Minimum passing score of 74%**

Who is responsible for the proper conduct and necessary supervision during an amateur operator license examination session?
- ✓ **Each administering VE**

What should a VE do if a candidate fails to comply with the examiner's instructions during an amateur operator license examination?
- ✓ **Immediately terminate the candidate's examination**

To which of the following examinees may a VE not administer an examination?
- ✓ **Relatives of the VE as listed in the FCC rules**

What may be the penalty for a VE who fraudulently administers or certifies an examination?
- ✓ **Revocation of the VE's amateur station license grant and the suspension of the VE's amateur operator license grant**

What must the administering VEs do after the administration of a successful examination for an amateur operator license?
- ✓ **They must submit the application document to the coordinating VEC according to the coordinating VEC instructions**

What must the VE team do if an examinee scores a passing grade on all examination elements needed for an upgrade or new license?
- ✓ **Three VEs must certify that the examinee is qualified for the license grant and that they have complied with the administering VE requirements**

What must the VE team do with the application form if the examinee does not pass the exam?

✓ **Return the application document to the examinee**

Which of these choices is an acceptable method for monitoring the applicants if a VEC opts to conduct an exam session remotely?

✓ **Use a real-time video link and the Internet to connect the exam session to the observing VEs**

For which types of out-of-pocket expenses do the Part 97 rules state that VEs and VECs may be reimbursed?

✓ **Preparing, processing, administering and coordinating an examination for an amateur radio license**

E1F – MISCELLANEOUS RULES: EXTERNAL RF POWER AMPLIFIERS; BUSINESS COMMUNIATIONS; COMPENSATED COMMUNICATIONS; SPREAD SPECTRUM; AUXILIARY STATIONS; RECIPROCAL OPERATING PRIVILEGES; SPECIAL TEMPORARY AUTHORITY

On what frequencies are spread spectrum transmissions permitted?

✓ **Only on amateur frequencies above 222 MHz**

What privileges are authorized in the U.S. to persons holding an amateur service license granted by the Government of Canada?

✓ **The operating terms and conditions of the Canadian amateur service license, not to exceed U.S. Extra Class privileges**

Under what circumstances may a dealer sell an external RF power amplifier capable of operation below 144 MHz if it has not been granted FCC certification?

✓ **It was purchased in used condition from an amateur operator and is sold to another amateur operator for use at that operator's station**

Which of the following geographic descriptions approximately describes "Line A"?

✓ **A line roughly parallel to and south of the U.S.-Canadian border**

Amateur stations may not transmit in which of the following frequency segments if they are located in the contiguous 48 states and north of Line A?

✓ **420 MHz - 430 MHz**

Under what circumstances might the FCC issue a Special Temporary Authority (STA) to an amateur station?

✓ **To provide for experimental amateur communications**

When may an amateur station send a message to a business?

✓ **When neither the amateur nor his or her employer has a pecuniary interest in the communications**

Which of the following types of amateur station communications are prohibited?
✓ **Communications transmitted for hire or material compensation, except as otherwise provided in the rules**

Which of the following conditions apply when transmitting spread spectrum emission?
A. A station transmitting SS emission must not cause harmful interference to other stations employing other authorized emissions
B. The transmitting station must be in an area regulated by the FCC or in a country that permits SS emissions
C. The transmission must not be used to obscure the meaning of any communication
✓ **D. All of these choices are correct**

What is the maximum permitted transmitter peak envelope power for an amateur station transmitting spread spectrum communications?
✓ **10 W**

Which of the following best describes one of the standards that must be met by an external RF power amplifier if it is to qualify for a grant of FCC certification?
✓ **It must satisfy the FCC's spurious emission standards when operated at the lesser of 1500 watts or its full output power**

Who may be the control operator of an auxiliary station?
✓ **Only Technician, General, Advanced or Amateur Extra Class operators**

Subelement E2 - Operating Procedures [5 Groups]

E2A - AMATEUR RADIO IN SPACE: AMATEUR SATELLITES; ORBITAL MECHANICS; FREQUENCIES AND MODES; SATELLITE HARDWARE; SATELLITE OPERATIONS; EXPERIMENTAL TELEMETRY APPLICATIONS

What is the direction of an ascending pass for an amateur satellite?
✓ **From south to north**

What is the direction of a descending pass for an amateur satellite?
✓ **From north to south**

What is the orbital period of an Earth satellite?
✓ **The time it takes for a satellite to complete one revolution around the Earth**

What is meant by the term mode as applied to an amateur radio satellite?
✓ **The satellite's uplink and downlink frequency bands**

What do the letters in a satellite's mode designator specify?
✓ **The uplink and downlink frequency ranges**

On what band would a satellite receive signals if it were operating in mode U/V?
✓ **435 MHz - 438 MHz**

Which of the following types of signals can be relayed through a linear transponder?
A. FM and CW
B. SSB and SSTV
C. PSK and Packet
✓ **D. All of these choices are correct**

Why should effective radiated power to a satellite which uses a linear transponder be limited?
✓ **To avoid reducing the downlink power to all other users**

What do the terms L band and S band specify with regard to satellite communications?
✓ **The 23 centimeter and 13 centimeter bands**

Why may the received signal from an amateur satellite exhibit a rapidly repeating fading effect?
✓ **Because the satellite is spinning**

What type of antenna can be used to minimize the effects of spin modulation and Faraday rotation?
✓ **A circularly polarized antenna**

What is one way to predict the location of a satellite at a given time?
✓ **By calculations using the Keplerian elements for the specified satellite**

What type of satellite appears to stay in one position in the sky?
✓ **Geostationary**

What technology is used to track, in real time, balloons carrying amateur radio transmitters?
✓ **APRS**

E2B - TELEVISION PRACTICES: FAST SCAN TELEVISION STANDARDS AND TECHNIQUES; SLOW SCAN TELEVISION STANDARDS AND TECHNIQUES

How many times per second is a new frame transmitted in a fast-scan (NTSC) television system?
✓ **30**

How many horizontal lines make up a fast-scan (NTSC) television frame?
✓ **525**

How is an interlaced scanning pattern generated in a fast-scan (NTSC) television system?
✓ **By scanning odd numbered lines in one field and even numbered lines in the next**

What is blanking in a video signal?
✓ **Turning off the scanning beam while it is traveling from right to left or from bottom to top**

Which of the following is an advantage of using vestigial sideband for standard fast-scan TV transmissions?
✓ **Vestigial sideband reduces bandwidth while allowing for simple video detector circuitry**

What is vestigial sideband modulation?
✓ **Amplitude modulation in which one complete sideband and a portion of the other are transmitted**

What is the name of the signal component that carries color information in NTSC video?
✓ **Chroma**

Which of the following is a common method of transmitting accompanying audio with amateur fast-scan television?
A. Frequency-modulated sub-carrier
B. A separate VHF or UHF audio link
C. Frequency modulation of the video carrier
✓ **D. All of these choices are correct**

What hardware, other than a receiver with SSB capability and a suitable computer, is needed to decode SSTV using Digital Radio Mondiale (DRM)?
✓ **No other hardware is needed**

Which of the following is an acceptable bandwidth for Digital Radio Mondiale (DRM) based voice or SSTV digital transmissions made on the HF amateur bands?
✓ **3 KHz**

What is the function of the Vertical Interval Signaling (VIS) code sent as part of an SSTV transmission?
✓ **To identify the SSTV mode being used**

How are analog SSTV images typically transmitted on the HF bands?
✓ **Varying tone frequencies representing the video are transmitted using single sideband**

How many lines are commonly used in each frame of an amateur slow-scan color television picture?
✓ **128 or 256**

What aspect of an amateur slow-scan television signal encodes the brightness of the picture?
✓ **Tone frequency**

What signals SSTV receiving equipment to begin a new picture line?
✓ **Specific tone frequencies**

Which is a video standard used by North American Fast Scan ATV stations?
✓ **NTSC**

What is the approximate bandwidth of a slow-scan TV signal?
✓ **3 kHz**

On which of the following frequencies is one likely to find FM ATV transmissions?
✓ **1255 MHz**

What special operating frequency restrictions are imposed on slow scan TV transmissions?
✓ **They are restricted to phone band segments and their bandwidth can be no greater than that of a voice signal of the same modulation type**

E2C – OPERATING METHODS: CONTEST AND DX OPERATING; REMOTE OPERATION TECHNIQUES; CABRILLO FORMAT; QSLING; RF NETWORK CONNECTED SYSTEMS

Which of the following is true about contest operating?
✓ **Operators are permitted to make contacts even if they do not submit a log**

Which of the following best describes the term self-spotting in regards to HF contest operating?
✓ **The generally prohibited practice of posting one's own call sign and frequency on a spotting network**

From which of the following bands is amateur radio contesting generally excluded?
✓ **30 m**

What type of transmission is most often used for a ham radio mesh network?
✓ **Spread spectrum in the 2.4 GHz band**

What is the function of a DX QSL Manager?
✓ **To handle the receiving and sending of confirmation cards for a DX station**

During a VHF/UHF contest, in which band segment would you expect to find the highest level of activity?
✓ **In the weak signal segment of the band, with most of the activity near the calling frequency**

What is the Cabrillo format?
✓ **A standard for submission of electronic contest logs**

Which of the following contacts may be confirmed through the U.S. QSL bureau system?
✓ **Contacts between a U.S. station and a non-U.S. station**

What type of equipment is commonly used to implement a ham radio mesh network?
✓ **A standard wireless router running custom software**

Why might a DX station state that they are listening on another frequency?
A. Because the DX station may be transmitting on a frequency that is prohibited to some responding stations
B. To separate the calling stations from the DX station
C. To improve operating efficiency by reducing interference
✓ **D. All of these choices are correct**

How should you generally identify your station when attempting to contact a DX station during a contest or in a pileup?
✓ **Send your full call sign once or twice**

What might help to restore contact when DX signals become too weak to copy across an entire HF band a few hours after sunset?
✓ **Switch to a lower frequency HF band**

What indicator is required to be used by U.S.-licensed operators when operating a station via remote control where the transmitter is located in the U.S.?
✓ **No additional indicator is required**

E2D - OPERATING METHODS: VHF AND UHF DIGITAL MODES AND PROCEDURES; APRS; EME PROCEDURES, METEOR SCATTER PROCEDURES

Which of the following digital modes is especially designed for use for meteor scatter signals?
✓ **FSK441**

Which of the following is a good technique for making meteor scatter contacts?
A. 15 second timed transmission sequences with stations alternating based on location
B. Use of high speed CW or digital modes
C. Short transmission with rapidly repeated call signs and signal reports
✓ **D. All of these choices are correct**

Which of the following digital modes is especially useful for EME communications?
✓ **JT65**

What is the purpose of digital store-and-forward functions on an Amateur Radio satellite?
✓ **To store digital messages in the satellite for later download by other stations**

Which of the following techniques is normally used by low Earth orbiting digital satellites to relay messages around the world?
✓ **Store-and-forward**

Which of the following describes a method of establishing EME contacts?
✓ **Time synchronous transmissions alternately from each station**

What digital protocol is used by APRS?
- ✓ **AX.25**

What type of packet frame is used to transmit APRS beacon data?
- ✓ **Unnumbered Information**

Which of these digital modes has the fastest data throughput under clear communication conditions?
- ✓ **300 baud packet**

How can an APRS station be used to help support a public service communications activity?
- ✓ **An APRS station with a GPS unit can automatically transmit information to show a mobile station's position during the event**

Which of the following data are used by the APRS network to communicate your location?
- ✓ **Latitude and longitude**

How does JT65 improve EME communications?
- ✓ **It can decode signals many dB below the noise floor using FEC**

What type of modulation is used for JT65 contacts?
- ✓ **Multi-tone AFSK**

What is one advantage of using JT65 coding?
- ✓ **The ability to decode signals which have a very low signal to noise ratio**

E2E – OPERATING METHODS: OPERATING HF DIGITAL MODES

Which type of modulation is common for data emissions below 30 MHz?
- ✓ **FSK**

What do the letters FEC mean as they relate to digital operation?
- ✓ **Forward Error Correction**

How is the timing of JT65 contacts organized?
- ✓ **Alternating transmissions at 1 minute intervals**

What is indicated when one of the ellipses in an FSK crossed-ellipse display suddenly disappears?
- ✓ **Selective fading has occurred**

Which type of digital mode does not support keyboard-to-keyboard operation?
- ✓ **Winlink**

What is the most common data rate used for HF packet?
- ✓ **300 baud**

What is the typical bandwidth of a properly modulated MFSK16 signal?
✓ **316 Hz**

Which of the following HF digital modes can be used to transfer binary files?
✓ **PACTOR**

Which of the following HF digital modes uses variable-length coding for bandwidth efficiency?
✓ **PSK31**

Which of these digital modes has the narrowest bandwidth?
✓ **PSK31**

What is the difference between direct FSK and audio FSK?
✓ **Direct FSK applies the data signal to the transmitter VFO**

Which type of control is used by stations using the Automatic Link Enable (ALE) protocol?
✓ **Automatic**

Which of the following is a possible reason that attempts to initiate contact with a digital station on a clear frequency are unsuccessful?
A. Your transmit frequency is incorrect
B. The protocol version you are using is not the supported by the digital station
C. Another station you are unable to hear is using the frequency
✓ **D. All of these choices are correct**

Subelement E3 - Radio Wave Propagation [3 Groups]

E3A - ELECTROMAGNETIC WAVES; EARTH-MOON-EARTH COMMUNICATIONS; METEOR SCATTER; MICROWAVE TROPOSPHERIC AND SCATTER PROPAGATION; AURORA PROPAGATION

What is the approximate maximum separation measured along the surface of the Earth between two stations communicating by Moon bounce?
✓ **12,000 miles, if the Moon is visible by both stations**

What characterizes libration fading of an EME signal?
✓ **A fluttery irregular fading**

When scheduling EME contacts, which of these conditions will generally result in the least path loss?
✓ **When the Moon is at perigee**

What do Hepburn maps predict?
✓ **Probability of tropospheric propagation**

Tropospheric propagation of microwave signals often occurs along what weather related structure?

✓ **Warm and cold fronts**

Which of the following is required for microwave propagation via rain scatter?

✓ **The rain must be within radio range of both stations**

Atmospheric ducts capable of propagating microwave signals often form over what geographic feature?

✓ **Bodies of water**

When a meteor strikes the Earth's atmosphere, a cylindrical region of free electrons is formed at what layer of the ionosphere?

✓ **The E layer**

Which of the following frequency range is most suited for meteor scatter communications?

✓ **28 MHz - 148 MHz**

Which type of atmospheric structure can create a path for microwave propagation?

✓ **Temperature inversion**

What is a typical range for tropospheric propagation of microwave signals?

✓ **100 miles to 300 miles**

What is the cause of auroral activity?

✓ **The interaction in the E layer of charged particles from the Sun with the Earth's magnetic field**

Which emission mode is best for aurora propagation?

✓ **CW**

From the contiguous 48 states, in which approximate direction should an antenna be pointed to take maximum advantage of aurora propagation?

✓ **North**

What is an electromagnetic wave?

✓ **A wave consisting of an electric field and a magnetic field oscillating at right angles to each other**

Which of the following best describes electromagnetic waves traveling in free space?

✓ **Changing electric and magnetic fields propagate the energy**

What is meant by circularly polarized electromagnetic waves?

✓ **WAVES WITH A ROTATING ELECTRIC FIELD**

E3B – TRANSEQUATORIAL PROPAGATION; LONG PATH; GRAY-LINE; MULTI-PATH; ORDINARY AND EXTRAORDINARY WAVES; CHORDAL HOP, SPORADIC E MECHANISMS

What is transequatorial propagation?
- ✓ **Propagation between two mid-latitude points at approximately the same distance north and south of the magnetic equator**

What is the approximate maximum range for signals using transequatorial propagation?
- ✓ **5000 miles**

What is the best time of day for transequatorial propagation?
- ✓ **Afternoon or early evening**

What is meant by the terms extraordinary and ordinary waves?
- ✓ **Independent waves created in the ionosphere that are elliptically polarized**

Which amateur bands typically support long-path propagation?
- ✓ **160 meters to 10 meters**

Which of the following amateur bands most frequently provides long-path propagation?
- ✓ **20 meters**

Which of the following could account for hearing an echo on the received signal of a distant station?
- ✓ **Receipt of a signal by more than one path**

What type of HF propagation is probably occurring if radio signals travel along the terminator between daylight and darkness?
- ✓ **Gray-line**

At what time of year is Sporadic E propagation most likely to occur?
- ✓ **Around the solstices, especially the summer solstice**

What is the cause of gray-line propagation?
- ✓ **At twilight and sunrise, D-layer absorption is low while E-layer and F-layer propagation remains high**

At what time of day is Sporadic-E propagation most likely to occur?
- ✓ **Any time**

What is the primary characteristic of chordal hop propagation?
- ✓ **Successive ionospheric reflections without an intermediate reflection from the ground**

Why is chordal hop propagation desirable?

✓ **The signal experiences less loss along the path compared to normal skip propagation**

What happens to linearly polarized radio waves that split into ordinary and extraordinary waves in the ionosphere?

✓ **They become elliptically polarized**

E3C – RADIO-PATH HORIZON; LESS COMMON PROPAGATION MODES; PROPAGATION PREDICTION TECHNIQUES AND MODELING; SPACE WEATHER PARAMETERS AND AMATEUR RADIO

What does the term ray tracing describe in regard to radio communications?

✓ **Modeling a radio wave's path through the ionosphere**

What is indicated by a rising A or K index?

✓ **Increasing disruption of the geomagnetic field**

Which of the following signal paths is most likely to experience high levels of absorption when the A index or K index is elevated?

✓ **Polar paths**

What does the value of Bz (B sub Z) represent?

✓ **Direction and strength of the interplanetary magnetic field**

What orientation of Bz (B sub z) increases the likelihood that incoming particles from the Sun will cause disturbed conditions?

✓ **Southward**

By how much does the VHF/UHF radio horizon distance exceed the geometric horizon?

✓ **By approximately 15 percent of the distance**

Which of the following descriptors indicates the greatest solar flare intensity?

✓ **Class X**

What does the space weather term G5 mean?

✓ **An extreme geomagnetic storm**

How does the intensity of an X3 flare compare to that of an X2 flare?

✓ **Twice as great**

What does the 304A solar parameter measure?

✓ **UV emissions at 304 angstroms, correlated to solar flux index**

What does VOACAP software model?

✓ **HF propagation**

How does the maximum distance of ground-wave propagation change when the signal frequency is increased?
✓ **It decreases**

What type of polarization is best for ground-wave propagation?
✓ **Vertical**

Why does the radio-path horizon distance exceed the geometric horizon?
✓ **Downward bending due to density variations in the atmosphere**

What might a sudden rise in radio background noise indicate?
✓ **A solar flare has occurred**

Subelement E4 - Amateur Practices [5 Groups]

E4A - TEST EQUIPMENT: ANALOG AND DIGITAL INSTRUMENTS; SPECTRUM AND NETWORK ANALYZERS, ANTENNA ANALYZERS; OSCILLOSCOPES; RF MEASUREMENTS; COMPUTER AIDED MEASUREMENTS

Which of the following parameter determines the bandwidth of a digital or computer-based oscilloscope?
✓ **Sampling rate**

Which of the following parameters would a spectrum analyzer display on the vertical and horizontal axes?
✓ **RF amplitude and frequency**

Which of the following test instrument is used to display spurious signals and/or intermodulation distortion products in an SSB transmitter?
✓ **A spectrum analyzer**

What determines the upper frequency limit for a computer soundcard-based oscilloscope program?
✓ **Analog-to-digital conversion speed of the soundcard**

What might be an advantage of a digital vs an analog oscilloscope?
A. Automatic amplitude and frequency numerical readout
B. Storage of traces for future reference
C. Manipulation of time base after trace capture
✓ **D. All of these choices are correct**

What is the effect of aliasing in a digital or computer-based oscilloscope?
✓ **False signals are displayed**

Which of the following is an advantage of using an antenna analyzer compared to an SWR bridge to measure antenna SWR?
✓ **Antenna analyzers do not need an external RF source**

Which of the following instrument would be best for measuring the SWR of a beam antenna?

✓ **An antenna analyzer**

When using a computer's soundcard input to digitize signals, what is the highest frequency signal that can be digitized without aliasing?

✓ **One-half the sample rate**

Which of the following displays multiple digital signal states simultaneously?

✓ **Logic analyzer**

Which of the following is good practice when using an oscilloscope probe?

✓ **Keep the signal ground connection of the probe as short as possible**

Which of the following procedures is an important precaution to follow when connecting a spectrum analyzer to a transmitter output?

✓ **Attenuate the transmitter output going to the spectrum analyzer**

How is the compensation of an oscilloscope probe typically adjusted?

✓ **A square wave is displayed and the probe is adjusted until the horizontal portions of the displayed wave are as nearly flat as possible**

What is the purpose of the prescaler function on a frequency counter?

✓ **It divides a higher frequency signal so a low-frequency counter can display the input frequency**

What is an advantage of a period-measuring frequency counter over a direct-count type?

✓ **It provides improved resolution of low-frequency signals within a comparable time period**

E4B – MEASUREMENT TECHNIQUE AND LIMITATIONS: INSTRUMENT ACCURACY AND PERFORMANCE LIMITATIONS; PROBES; TECHNICQUES TO MINIMIZE ERRORS; MEASUREMENT OF "Q"; INSTRUMENT CALIBRATION; S PARAMETERS; VECTOR NETWORK ANALYZERS

Which of the following factors most affects the accuracy of a frequency counter?

✓ **Time base accuracy**

What is an advantage of using a bridge circuit to measure impedance?

✓ **It is very precise in obtaining a signal null**

If a frequency counter with a specified accuracy of +/- 1.0 ppm reads 146,520,000 Hz, what is the most the actual frequency being measured could differ from the reading?

✓ **146.52 Hz**

If a frequency counter with a specified accuracy of +/- 0.1 ppm reads 146,520,000 Hz, what is the most the actual frequency being measured could differ from the reading?

✓ **14.652 Hz**

If a frequency counter with a specified accuracy of +/- 10 ppm reads 146,520,000 Hz, what is the most the actual frequency being measured could differ from the reading?

✓ **1465.20 Hz**

How much power is being absorbed by the load when a directional power meter connected between a transmitter and a terminating load reads 100 watts forward power and 25 watts reflected power?

✓ **75 watts**

What do the subscripts of S parameters represent?

✓ **The port or ports at which measurements are made**

Which of the following is a characteristic of a good DC voltmeter?

✓ **High impedance input**

What is indicated if the current reading on an RF ammeter placed in series with the antenna feed line of a transmitter increases as the transmitter is tuned to resonance?

✓ **There is more power going into the antenna**

Which of the following describes a method to measure intermodulation distortion in an SSB transmitter?

✓ **Modulate the transmitter with two non-harmonically related audio frequencies and observe the RF output with a spectrum analyzer**

How should an antenna analyzer be connected when measuring antenna resonance and feed point impedance?

✓ **Connect the antenna feed line directly to the analyzer's connector**

What is the significance of voltmeter sensitivity expressed in ohms per volt?

✓ **The full scale reading of the voltmeter multiplied by its ohms per volt rating will indicate the input impedance of the voltmeter**

Which S parameter is equivalent to forward gain?

✓ **S21**

What happens if a dip meter is too tightly coupled to a tuned circuit being checked?

✓ **A less accurate reading results**

Which of the following can be used as a relative measurement of the Q for a series-tuned circuit?

✓ **The bandwidth of the circuit's frequency response**

Which S parameter represents return loss or SWR?

✓ **S11**

What three test loads are used to calibrate a standard RF vector network analyzer?
- ✓ **Short circuit, open circuit, and 50 ohms**

E4C - RECEIVER PERFORMANCE CHARACTERISTICS, PHASE NOISE, NOISE FLOOR, IMAGE REJECTION, MDS, SIGNAL-TO-NOISE-RATIO; SELECTIVITY; EFFECTS OF SDR RECEIVER NON-LINEARITY

What is an effect of excessive phase noise in the local oscillator section of a receiver?
- ✓ **It can cause strong signals on nearby frequencies to interfere with reception of weak signals**

Which of the following portions of a receiver can be effective in eliminating image signal interference?
- ✓ **A front-end filter or pre-selector**

What is the term for the blocking of one FM phone signal by another, stronger FM phone signal?
- ✓ **Capture effect**

How is the noise figure of a receiver defined?
- ✓ **The ratio in dB of the noise generated by the receiver to the theoretical minimum noise**

What does a value of -174 dBm/Hz represent with regard to the noise floor of a receiver?
- ✓ **The theoretical noise at the input of a perfect receiver at room temperature**

A CW receiver with the AGC off has an equivalent input noise power density of -174 dBm/Hz. What would be the level of an unmodulated carrier input to this receiver that would yield an audio output SNR of 0 dB in a 400 Hz noise bandwidth?
- ✓ **-148 dBm**

What does the MDS of a receiver represent?
- ✓ **The minimum discernible signal**

An SDR receiver is overloaded when input signals exceed what level?
- ✓ **The maximum count value of the analog-to-digital converter**

Which of the following choices is a good reason for selecting a high frequency for the design of the IF in a conventional HF or VHF communications receiver?
- ✓ **Easier for front-end circuitry to eliminate image responses**

Which of the following is a desirable amount of selectivity for an amateur RTTY HF receiver?
- ✓ **300 Hz**

Which of the following is a desirable amount of selectivity for an amateur SSB phone receiver?

✓ **2.4 kHz**

What is an undesirable effect of using too wide a filter bandwidth in the IF section of a receiver?

✓ **Undesired signals may be heard**

How does a narrow-band roofing filter affect receiver performance?

✓ **It improves dynamic range by attenuating strong signals near the receive frequency**

What transmit frequency might generate an image response signal in a receiver tuned to 14.300 MHz and which uses a 455 kHz IF frequency?

✓ **15.210 MHz**

What is usually the primary source of noise that is heard from an HF receiver with an antenna connected?

✓ **Atmospheric noise**

Which of the following is caused by missing codes in an SDR receiver's analog-to-digital converter?

✓ **Distortion**

Which of the following has the largest effect on an SDR receiver's linearity?

✓ **Analog-to-digital converter sample width in bits**

E4D - RECEIVER PERFORMANCE CHARACTERISTICS: BLOCKING DYNAMIC RANGE; INTERMODULATION AND CROSS-MODULATION INTERFERENCE; 3RD ORDER INTERCEPT; DESENSITIZATION; PRESELECTOR

What is meant by the blocking dynamic range of a receiver?

✓ **The difference in dB between the noise floor and the level of an incoming signal which will cause 1 dB of gain compression**

Which of the following describes two problems caused by poor dynamic range in a communications receiver?

✓ **Cross-modulation of the desired signal and desensitization from strong adjacent signals**

How can intermodulation interference between two repeaters occur?

✓ **When the repeaters are in close proximity and the signals mix in the final amplifier of one or both transmitters**

Which of the following may reduce or eliminate intermodulation interference in a repeater caused by another transmitter operating in close proximity?

✓ **A properly terminated circulator at the output of the transmitter**

What transmitter frequencies would cause an intermodulation-product signal in a receiver tuned to 146.70 MHz when a nearby station transmits on 146.52 MHz?
✓ **146.34 MHz and 146.61 MHz**

What is the term for unwanted signals generated by the mixing of two or more signals?
✓ **Intermodulation interference**

Which describes the most significant effect of an off-frequency signal when it is causing cross-modulation interference to a desired signal?
✓ **The off-frequency unwanted signal is heard in addition to the desired signal**

What causes intermodulation in an electronic circuit?
✓ **Nonlinear circuits or devices**

What is the purpose of the preselector in a communications receiver?
✓ **To increase rejection of unwanted signals**

What does a third-order intercept level of 40 dBm mean with respect to receiver performance?
✓ **A pair of 40 dBm signals will theoretically generate a third-order intermodulation product with the same level as the input signals**

Why are third-order intermodulation products created within a receiver of particular interest compared to other products?
✓ **The third-order product of two signals which are in the band of interest is also likely to be within the band**

What is the term for the reduction in receiver sensitivity caused by a strong signal near the received frequency?
✓ **Desensitization**

Which of the following can cause receiver desensitization?
✓ **Strong adjacent channel signals**

Which of the following is a way to reduce the likelihood of receiver desensitization?
✓ **Decrease the RF bandwidth of the receiver**

E4E - NOISE SUPPRESSION: SYSTEM NOISE; ELECTRICAL APPLIANCE NOISE; LINE NOISE; LOCATING NOISE SOURCES; DSP NOISE REDUCTION; NOISE BLANKERS; GROUNDING FOR SIGNALS

Which of the following types of receiver noise can often be reduced by use of a receiver noise blanker?
✓ **Ignition noise**

Which of the following types of receiver noise can often be reduced with a DSP noise filter?
A. Broadband white noise
B. Ignition noise
C. Power line noise
✓ **D. All of these choices are correct**

Which of the following signals might a receiver noise blanker be able to remove from desired signals?
✓ **Signals which appear across a wide bandwidth**

How can conducted and radiated noise caused by an automobile alternator be suppressed?
✓ **By connecting the radio's power leads directly to the battery and by installing coaxial capacitors in line with the alternator leads**

How can noise from an electric motor be suppressed?
✓ **By installing a brute-force AC-line filter in series with the motor leads**

What is a major cause of atmospheric static?
✓ **Thunderstorms**

How can you determine if line noise interference is being generated within your home?
✓ **By turning off the AC power line main circuit breaker and listening on a battery operated radio**

What type of signal is picked up by electrical wiring near a radio antenna?
✓ **A common-mode signal at the frequency of the radio transmitter**

What undesirable effect can occur when using an IF noise blanker?
✓ **Nearby signals may appear to be excessively wide even if they meet emission standards**

What is a common characteristic of interference caused by a touch controlled electrical device?
A. The interfering signal sounds like AC hum on an AM receiver or a carrier modulated by 60 Hz hum on a SSB or CW receiver
B. The interfering signal may drift slowly across the HF spectrum
C. The interfering signal can be several kHz in width and usually repeats at regular intervals across a HF band
✓ **D. All of these choices are correct**

Which is the most likely cause if you are hearing combinations of local AM broadcast signals within one or more of the MF or HF ham bands?
✓ **Nearby corroded metal joints are mixing and re-radiating the broadcast signals**

What is one disadvantage of using some types of automatic DSP notch-filters when attempting to copy CW signals?

✓ **A DSP filter can remove the desired signal at the same time as it removes interfering signals**

What might be the cause of a loud roaring or buzzing AC line interference that comes and goes at intervals?

A. Arcing contacts in a thermostatically controlled device
B. A defective doorbell or doorbell transformer inside a nearby residence
C. A malfunctioning illuminated advertising display
✓ **D. All of these choices are correct**

What is one type of electrical interference that might be caused by the operation of a nearby personal computer?

✓ **The appearance of unstable modulated or unmodulated signals at specific frequencies**

Which of the following can cause shielded cables to radiate or receive interference?

✓ **Common mode currents on the shield and conductors**

What current flows equally on all conductors of an unshielded multi-conductor cable?

✓ **Common-mode current**

Subelement E5 - Electrical Principles [4 Groups]

E5A - RESONANCE AND Q: CHARACTERISTICS OF RESONANT CIRCUITS: SERIES AND PARALLEL RESONANCE; DEFINITIONS AND EFFECTS OF Q; HALF-POWER BANDWIDTH; PHASE RELATIONSHIPS IN REACTIVE CIRCUITS

What can cause the voltage across reactances in series to be larger than the voltage applied to them?

✓ **Resonance**

What is resonance in an electrical circuit?

✓ **The frequency at which the capacitive reactance equals the inductive reactance**

What is the magnitude of the impedance of a series RLC circuit at resonance?

✓ **Approximately equal to circuit resistance**

What is the magnitude of the impedance of a circuit with a resistor, an inductor and a capacitor all in parallel, at resonance?

✓ **Approximately equal to circuit resistance**

What is the magnitude of the current at the input of a series RLC circuit as the frequency goes through resonance?

✓ **Maximum**

What is the magnitude of the circulating current within the components of a parallel LC circuit at resonance?

✓ **It is at a maximum**

What is the magnitude of the current at the input of a parallel RLC circuit at resonance?

✓ **Minimum**

What is the phase relationship between the current through and the voltage across a series resonant circuit at resonance?

✓ **The voltage and current are in phase**

How is the Q of an RLC parallel resonant circuit calculated?

✓ **Resistance divided by the reactance of either the inductance or capacitance**

How is the Q of an RLC series resonant circuit calculated?

✓ **Reactance of either the inductance or capacitance divided by the resistance**

What is the half-power bandwidth of a parallel resonant circuit that has a resonant frequency of 7.1 MHz and a Q of 150?

✓ **47.3 kHz**

What is the half-power bandwidth of a parallel resonant circuit that has a resonant frequency of 3.7 MHz and a Q of 118?

✓ **31.4 kHz**

What is an effect of increasing Q in a resonant circuit?

✓ **Internal voltages and circulating currents increase**

What is the resonant frequency of a series RLC circuit if R is 22 ohms, L is 50 microhenrys and C is 40 picofarads?

✓ **3.56 MHz**

Which of the following can increase Q for inductors and capacitors?

✓ **Lower losses**

What is the resonant frequency of a parallel RLC circuit if R is 33 ohms, L is 50 microhenrys and C is 10 picofarads?

✓ **7.12 MHz**

What is the result of increasing the Q of an impedance-matching circuit?

✓ **Matching bandwidth is decreased**

What is the term for the time required for the capacitor in an RC circuit to be charged to 63.2% of the applied voltage?
✓ **One time constant**

What is the term for the time it takes for a charged capacitor in an RC circuit to discharge to 36.8% of its initial voltage?
✓ **One time constant**

What happens to the phase angle of a reactance when it is converted to a susceptance?
✓ **The sign is reversed**

What is the time constant of a circuit having two 220 microfarad capacitors and two 1 megohm resistors, all in parallel?
✓ **220 seconds**

What happens to the magnitude of a reactance when it is converted to a susceptance?
✓ **The magnitude of the susceptance is the reciprocal of the magnitude of the reactance**

What is susceptance?
✓ **The inverse of reactance**

What is the phase angle between the voltage across and the current through a series RLC circuit if XC is 500 ohms, R is 1 kilohm, and XL is 250 ohms?
✓ **14.0 degrees with the voltage lagging the current**

What is the phase angle between the voltage across and the current through a series RLC circuit if XC is 100 ohms, R is 100 ohms, and XL is 75 ohms?
✓ **14 degrees with the voltage lagging the current**

What is the relationship between the current through a capacitor and the voltage across a capacitor?
✓ **Current leads voltage by 90 degrees**

What is the relationship between the current through an inductor and the voltage across an inductor?
✓ **Voltage leads current by 90 degrees**

What is the phase angle between the voltage across and the current through a series RLC circuit if XC is 25 ohms, R is 100 ohms, and XL is 50 ohms?
✓ **14 degrees with the voltage leading the current**

What is admittance?
✓ **The inverse of impedance**
What letter is commonly used to represent susceptance?
✓ **B**

E5C - COORDINATE SYSTEMS AND PHASORS IN ELECTRONICS: RECTANGULAR
    COORDINATES; POLAR COORDINATES; PHASORS

Which of the following represents a capacitive reactance in rectangular notation?
✓ **–jX**

How are impedances described in polar coordinates?
✓ **By phase angle and amplitude**

Which of the following represents an inductive reactance in polar coordinates?
✓ **A positive phase angle**

Which of the following represents a capacitive reactance in polar coordinates?
✓ **A negative phase angle**

What is the name of the diagram used to show the phase relationship between
impedances at a given frequency?
✓ **Phasor diagram**

What does the impedance 50–j25 represent?
✓ **50 ohms resistance in series with 25 ohms capacitive reactance**

What is a vector?
✓ **A quantity with both magnitude and an angular component**

What coordinate system is often used to display the phase angle of a circuit containing
resistance, inductive and/or capacitive reactance?
✓ **Polar coordinates**

When using rectangular coordinates to graph the impedance of a circuit, what does the
horizontal axis represent?
✓ **Resistive component**

When using rectangular coordinates to graph the impedance of a circuit, what does the
vertical axis represent?
✓ **Reactive component**

What do the two numbers that are used to define a point on a graph using rectangular
coordinates represent?
✓ **The coordinate values along the horizontal and vertical axes**

If you plot the impedance of a circuit using the rectangular coordinate system and find the impedance point falls on the right side of the graph on the horizontal axis, what do you know about the circuit?
✓ **It is equivalent to a pure resistance**

What coordinate system is often used to display the resistive, inductive, and/or capacitive reactance components of impedance?
✓ **Rectangular coordinates**

Which point on Figure E5-2 best represents the impedance of a series circuit consisting of a 400 ohm resistor and a 38 picofarad capacitor at 14 MHz?
✓ **Point 4**

Which point in Figure E5-2 best represents the impedance of a series circuit consisting of a 300 ohm resistor and an 18 microhenry inductor at 3.505 MHz?
✓ **Point 3**

Which point on Figure E5-2 best represents the impedance of a series circuit consisting of a 300 ohm resistor and a 19 picofarad capacitor at 21.200 MHz?
✓ **Point 1**

Which point on Figure E5-2 best represents the impedance of a series circuit consisting of a 300 ohm resistor, a 0.64-microhenry inductor and an 85-picofarad capacitor at 24.900 MHz?
✓ **Point 8**

E5D - AC AND RF ENERGY IN REAL CIRCUITS: SKIN EFFECT; ELECTROSTATIC AND ELECTROMAGNETIC FIELDS; REACTIVE POWER; POWER FACTOR; ELECTRICAL LENGTH OF CONDUCTORS AT UHF AND MICROWAVE FREQUENCIES

What is the result of skin effect?
✓ **As frequency increases, RF current flows in a thinner layer of the conductor, closer to the surface**

Why is it important to keep lead lengths short for components used in circuits for VHF and above?
✓ **To avoid unwanted inductive reactance**

What is microstrip?
✓ **Precision printed circuit conductors above a ground plane that provide constant impedance interconnects at microwave frequencies**

Why are short connections necessary at microwave frequencies?
✓ **To reduce phase shift along the connection**

Which parasitic characteristic increases with conductor length?
✓ **Inductance**

In what direction is the magnetic field oriented about a conductor in relation to the direction of electron flow?
✓ **In a direction determined by the left-hand rule**

What determines the strength of the magnetic field around a conductor?
✓ **The amount of current flowing through the conductor**

What type of energy is stored in an electromagnetic or electrostatic field?
✓ **Potential energy**

What happens to reactive power in an AC circuit that has both ideal inductors and ideal capacitors?
✓ **It is repeatedly exchanged between the associated magnetic and electric fields, but is not dissipated**

How can the true power be determined in an AC circuit where the voltage and current are out of phase?
✓ **By multiplying the apparent power times the power factor**

What is the power factor of an R-L circuit having a 60 degree phase angle between the voltage and the current?
✓ **0.5**

How many watts are consumed in a circuit having a power factor of 0.2 if the input is 100-VAC at 4 amperes?
✓ **80 watts**

How much power is consumed in a circuit consisting of a 100 ohm resistor in series with a 100 ohm inductive reactance drawing 1 ampere?
✓ **100 Watts**

What is reactive power?
✓ **Wattless, nonproductive power**

What is the power factor of an R-L circuit having a 45 degree phase angle between the voltage and the current?
✓ **0.707**

What is the power factor of an R-L circuit having a 30 degree phase angle between the voltage and the current?
✓ **0.866**

How many watts are consumed in a circuit having a power factor of 0.6 if the input is 200VAC at 5 amperes?
✓ **600 watts**

How many watts are consumed in a circuit having a power factor of 0.71 if the apparent power is 500VA?

✓ **355 W**

Subelement E6 - Circuit Components [6 Groups]

<u>E6A - SEMICONDUCTOR MATERIALS AND DEVICES: SEMICONDUCTOR MATERIALS; GERMANIUM, SILICON, P-TYPE, N-TYPE; TRANSISTOR TYPES: NPN, PNP, JUNCTION, FIELD-EFFECT TRANSISTORS: ENHANCEMENT MODE; DEPLETION MODE; MOS; CMOS; N-CHANNEL; P-CHANNEL</u>

In what application is gallium arsenide used as a semiconductor material in preference to germanium or silicon?

✓ **In microwave circuits**

Which of the following semiconductor materials contains excess free electrons?

✓ **N-type**

Why does a PN-junction diode not conduct current when reverse biased?

✓ **Holes in P-type material and electrons in the N-type material are separated by the applied voltage, widening the depletion region**

What is the name given to an impurity atom that adds holes to a semiconductor crystal structure?

✓ **Acceptor impurity**

What is the alpha of a bipolar junction transistor?

✓ **The change of collector current with respect to emitter current**

What is the beta of a bipolar junction transistor?

✓ **The change in collector current with respect to base current**

Which of the following indicates that a silicon NPN junction transistor is biased on?

✓ **Base-to-emitter voltage of approximately 0.6 to 0.7 volts**

What term indicates the frequency at which the grounded-base current gain of a transistor has decreased to 0.7 of the gain obtainable at 1 kHz?

✓ **Alpha cutoff frequency**

What is a depletion-mode FET?

✓ **An FET that exhibits a current flow between source and drain when no gate voltage is applied**

In Figure E6-2, what is the schematic symbol for an N-channel dual-gate MOSFET?

✓ **4**

In Figure E6-2, what is the schematic symbol for a P-channel junction FET?

✓ **1**

Why do many MOSFET devices have internally connected Zener diodes on the gates?
- ✓ **To reduce the chance of the gate insulation being punctured by static discharges or excessive voltages**

What do the initials CMOS stand for?
- ✓ **Complementary Metal-Oxide Semiconductor**

How does DC input impedance at the gate of a field-effect transistor compare with the DC input impedance of a bipolar transistor?
- ✓ **An FET has high input impedance; a bipolar transistor has low input impedance**

Which semiconductor material contains excess holes in the outer shell of electrons?
- ✓ **P-type**

What are the majority charge carriers in N-type semiconductor material?
- ✓ **Free electrons**

What are the names of the three terminals of a field-effect transistor?
- ✓ **Gate, drain, source**

E6B - DIODES

What is the most useful characteristic of a Zener diode?
- ✓ **A constant voltage drop under conditions of varying current**

What is an important characteristic of a Schottky diode as compared to an ordinary silicon diode when used as a power supply rectifier?
- ✓ **Less forward voltage drop**

What special type of diode is capable of both amplification and oscillation?
- ✓ **Tunnel**

What type of semiconductor device is designed for use as a voltage-controlled capacitor?
- ✓ **Varactor diode**

What characteristic of a PIN diode makes it useful as an RF switch or attenuator?
- ✓ **A large region of intrinsic material**

Which of the following is a common use of a hot-carrier diode?
- ✓ **As a VHF/UHF mixer or detector**

What is the failure mechanism when a junction diode fails due to excessive current?
- ✓ **Excessive junction temperature**

Which of the following describes a type of semiconductor diode?
- ✓ **Metal-semiconductor junction**

What is a common use for point contact diodes?
✓ **As an RF detector**

In Figure E6-3, what is the schematic symbol for a light-emitting diode?
✓ **5**

What is used to control the attenuation of RF signals by a PIN diode?
✓ **Forward DC bias current**

What is one common use for PIN diodes?
✓ **As an RF switch**

What type of bias is required for an LED to emit light?
✓ **Forward bias**

E6C - DIGITAL ICS: FAMILIES OF DIGITAL ICS; GATES; PROGRAMMABLE LOGIC DEVICES (PLDS)

What is the function of hysteresis in a comparator?
✓ **To prevent input noise from causing unstable output signals**

What happens when the level of a comparator's input signal crosses the threshold?
✓ **The comparator changes its output state**

What is tri-state logic?
✓ **Logic devices with 0, 1, and high impedance output states**

What is the primary advantage of tri-state logic?
✓ **Ability to connect many device outputs to a common bus**

What is an advantage of CMOS logic devices over TTL devices?
✓ **Lower power consumption**

Why do CMOS digital integrated circuits have high immunity to noise on the input signal or power supply?
✓ **The input switching threshold is about one-half the power supply voltage**

What best describes a pull-up or pull-down resistor?
✓ **A resistor connected to the positive or negative supply line used to establish a voltage when an input or output is an open circuit**

In Figure E6-5, what is the schematic symbol for a NAND gate?
✓ **2**

What is a Programmable Logic Device (PLD)?
✓ **A programmable collection of logic gates and circuits in a single integrated circuit**

In Figure E6-5, what is the schematic symbol for a NOR gate?
✓ **4**

In Figure E6-5, what is the schematic symbol for the NOT operation (inverter)?
✓ **5**

What is BiCMOS logic?
✓ **An integrated circuit logic family using both bipolar and CMOS transistors**

Which of the following is an advantage of BiCMOS logic?
✓ **It has the high input impedance of CMOS and the low output impedance of bipolar transistors**

What is the primary advantage of using a Programmable Gate Array (PGA) in a logic circuit?
✓ **Complex logic functions can be created in a single integrated circuit**

E6D - TOROIDAL AND SOLENOIDAL INDUCTORS: PERMEABILITY, CORE MATERIAL, SELECTING, WINDING; TRANSFORMERS; PIEZOELECTRIC DEVICES

How many turns will be required to produce a 5-microhenry inductor using a powdered-iron toroidal core that has an inductance index (A L) value of 40 microhenrys/100 turns?
✓ **35 turns**

What is the equivalent circuit of a quartz crystal?
✓ **Motional capacitance, motional inductance, and loss resistance in series, all in parallel with a shunt capacitor representing electrode and stray capacitance**

Which of the following is an aspect of the piezoelectric effect?
✓ **Mechanical deformation of material by the application of a voltage**

Which materials are commonly used as a slug core in a variable inductor?
✓ **Ferrite and brass**

What is one reason for using ferrite cores rather than powdered-iron in an inductor?
✓ **Ferrite toroids generally require fewer turns to produce a given inductance value**

What core material property determines the inductance of a toroidal inductor?
✓ **Permeability**

What is the usable frequency range of inductors that use toroidal cores, assuming a correct selection of core material for the frequency being used?
✓ **From less than 20 Hz to approximately 300 MHz**

What is one reason for using powdered-iron cores rather than ferrite cores in an inductor?

✓ **Powdered-iron cores generally maintain their characteristics at higher currents**

What devices are commonly used as VHF and UHF parasitic suppressors at the input and output terminals of a transistor HF amplifier?

✓ **Ferrite beads**

What is a primary advantage of using a toroidal core instead of a solenoidal core in an inductor?

✓ **Toroidal cores confine most of the magnetic field within the core material**

How many turns will be required to produce a 1-mH inductor using a core that has an inductance index (A L) value of 523 millihenrys/1000 turns?

✓ **43 turns**

What is the definition of saturation in a ferrite core inductor?

✓ **The ability of the inductor's core to store magnetic energy has been exceeded**

What is the primary cause of inductor self-resonance?

✓ **Inter-turn capacitance**

Which type of slug material decreases inductance when inserted into a coil?

✓ **Brass**

What is current in the primary winding of a transformer called if no load is attached to the secondary?

✓ **Magnetizing current**

What is the common name for a capacitor connected across a transformer secondary that is used to absorb transient voltage spikes?

✓ **Snubber capacitor**

Why should core saturation of a conventional impedance matching transformer be avoided?

✓ **Harmonics and distortion could result**

E6E - ANALOG ICS: MMICS, CCDS, DEVICE PACKAGES

Which of the following is true of a charge-coupled device (CCD)?

✓ **It samples an analog signal and passes it in stages from the input to the output**

Which of the following device packages is a through-hole type?

✓ **DIP**

Which of the following materials is likely to provide the highest frequency of operation when used in MMICs?

✓ **Gallium nitride**

Which is the most common input and output impedance of circuits that use MMICs?
✓ **50 ohms**

Which of the following noise figure values is typical of a low-noise UHF preamplifier?
✓ **2 dB**

What characteristics of the MMIC make it a popular choice for VHF through microwave circuits?
✓ **Controlled gain, low noise figure, and constant input and output impedance over the specified frequency range**

Which of the following is typically used to construct a MMIC-based microwave amplifier?
✓ **Microstrip construction**

How is voltage from a power supply normally furnished to the most common type of monolithic microwave integrated circuit (MMIC)?
✓ **Through a resistor and/or RF choke connected to the amplifier output lead**

Which of the following component package types would be most suitable for use at frequencies above the HF range?
✓ **Surface mount**

What is the packaging technique in which leadless components are soldered directly to circuit boards?
✓ **Surface mount**

What is a characteristic of DIP packaging used for integrated circuits?
✓ **A total of two rows of connecting pins placed on opposite sides of the package (Dual In-line Package)**

Why are high-power RF amplifier ICs and transistors sometimes mounted in ceramic packages?
✓ **Better dissipation of heat**

E6F - OPTICAL COMPONENTS: PHOTOCONDUCTIVE PRINCIPLES AND EFFECTS, PHOTOVOLTAIC SYSTEMS, OPTICAL COUPLERS, OPTICAL SENSORS, AND OPTOISOLATORS; LCDS

What is photoconductivity?
✓ **The increased conductivity of an illuminated semiconductor**

What happens to the conductivity of a photoconductive material when light shines on it?
✓ **It increases**

What is the most common configuration of an optoisolator or optocoupler?
✓ **An LED and a phototransistor**

What is the photovoltaic effect?
✓ **The conversion of light to electrical energy**

Which describes an optical shaft encoder?
✓ **A device which detects rotation of a control by interrupting a light source with a patterned wheel**

Which of these materials is affected the most by photoconductivity?
✓ **A crystalline semiconductor**

What is a solid state relay?
✓ **A device that uses semiconductors to implement the functions of an electromechanical relay**

Why are optoisolators often used in conjunction with solid state circuits when switching 120VAC?
✓ **Optoisolators provide a very high degree of electrical isolation between a control circuit and the circuit being switched**

What is the efficiency of a photovoltaic cell?
✓ **The relative fraction of light that is converted to current**

What is the most common type of photovoltaic cell used for electrical power generation?
✓ **Silicon**

What is the approximate open-circuit voltage produced by a fully-illuminated silicon photovoltaic cell?
✓ **0.5 V**

What absorbs the energy from light falling on a photovoltaic cell?
✓ **Electrons**

What is a liquid crystal display (LCD)?
✓ **A display utilizing a crystalline liquid and polarizing filters which becomes opaque when voltage is applied**

Which of the following is true of LCD displays?
✓ **They may be hard view through polarized lenses**

Subelement E7 - Practical Circuits [8 Groups]

E7A - DIGITAL CIRCUITS: DIGITAL CIRCUIT PRINCIPLES AND LOGIC CIRCUITS: CLASSES OF LOGIC ELEMENTS; POSITIVE AND NEGATIVE LOGIC; FREQUENCY DIVIDERS; TRUTH TABLES

Which is a bi-stable circuit?
- ✓ **A flip-flop**

What is the function of a decade counter digital IC?
- ✓ **It produces one output pulse for every ten input pulses**

Which of the following can divide the frequency of a pulse train by 2?
- ✓ **A flip-flop**

How many flip-flops are required to divide a signal frequency by 4?
- ✓ **2**

Which of the following is a circuit that continuously alternates between two states without an external clock?
- ✓ **Astable multivibrator**

What is a characteristic of a monostable multivibrator?
- ✓ **It switches momentarily to the opposite binary state and then returns to its original state after a set time**

What logical operation does a NAND gate perform?
- ✓ **It produces logic "0" at its output only when all inputs are logic "1"**

What logical operation does an OR gate perform?
- ✓ **It produces logic "1" at its output if any or all inputs are logic "1"**

What logical operation is performed by an exclusive NOR gate?
- ✓ **It produces logic "0" at its output if any single input is logic "1"**

What is a truth table?
- ✓ **A list of inputs and corresponding outputs for a digital device**

What type of logic defines "1" as a high voltage?
- ✓ **Positive Logic**

What type of logic defines "0" as a high voltage?
- ✓ **Positive Logic**

For what portion of a signal cycle does a Class AB amplifier operate?
✓ **More than 180 degrees but less than 360 degrees**

What is a Class D amplifier?
✓ **A type of amplifier that uses switching technology to achieve high efficiency**

Which of the following components form the output of a class D amplifier circuit?
✓ **A low-pass filter to remove switching signal components**

Where on the load line of a Class A common emitter amplifier would bias normally be set?
✓ **Approximately half-way between saturation and cutoff**

What can be done to prevent unwanted oscillations in an RF power amplifier?
✓ **Install parasitic suppressors and/or neutralize the stage**

Which of the following amplifier types reduces or eliminates even order harmonics?
✓ **Push-pull**

Which of the following is a likely result when a Class C amplifier is used to amplify a single-sideband phone signal?
✓ **Signal distortion and excessive bandwidth**

How can an RF power amplifier be neutralized?
✓ **By feeding a 180-degree out-of-phase portion of the output back to the input**

Which of the following describes how the loading and tuning capacitors are to be adjusted when tuning a vacuum tube RF power amplifier that employs a Pi-network output circuit?
✓ **The tuning capacitor is adjusted for minimum plate current, and the loading capacitor is adjusted for maximum permissible plate current**

In Figure E7-1, what is the purpose of R1 and R2?
✓ **Fixed bias**

In Figure E7-1, what is the purpose of R3?
✓ **Self bias**

What type of amplifier circuit is shown in Figure E7-1?
✓ **Common emitter**

In Figure E7-2, what is the purpose of R?
✓ **Emitter load**

Why are switching amplifiers more efficient than linear amplifiers?
✓ **The power transistor is at saturation or cut off most of the time, resulting in low power dissipation**

What is one way to prevent thermal runaway in a bipolar transistor amplifier?
✓ **Use a resistor in series with the emitter**

What is the effect of intermodulation products in a linear power amplifier?
✓ **Transmission of spurious signals**

Why are odd-order rather than even-order intermodulation distortion products of concern in linear power amplifiers?
✓ **Because they are relatively close in frequency to the desired signal**

What is a characteristic of a grounded-grid amplifier?
✓ **Low input impedance**

E7C - FILTERS AND MATCHING NETWORKS: TYPES OF NETWORKS; TYPES OF FILTERS; FILTER APPLICATIONS; FILTER CHARACTERISTICS; IMPEDANCE MATCHING; DSP FILTERING

How are the capacitors and inductors of a low-pass filter Pi-network arranged between the network's input and output?
✓ **A capacitor is connected between the input and ground, another capacitor is connected between the output and ground, and an inductor is connected between input and output**

Which of the following is a property of a T-network with series capacitors and a parallel shunt inductor?
✓ **It is a high-pass filter**

What advantage does a Pi-L-network have over a regular Pi-network for impedance matching between the final amplifier of a vacuum-tube transmitter and an antenna?
✓ **Greater harmonic suppression**

How does an impedance-matching circuit transform a complex impedance to a resistive impedance?
✓ **It cancels the reactive part of the impedance and changes the resistive part to a desired value**

Which filter type is described as having ripple in the passband and a sharp cutoff?
✓ **A Chebyshev filter**

What are the distinguishing features of an elliptical filter?
✓ **Extremely sharp cutoff with one or more notches in the stop band**

What kind of filter would you use to attenuate an interfering carrier signal while receiving an SSB transmission?
✓ **A notch filter**

Which of the following factors has the greatest effect in helping determine the bandwidth and response shape of a crystal ladder filter?
✓ **The relative frequencies of the individual crystals**

What is a Jones filter as used as part of an HF receiver IF stage?
✓ **A variable bandwidth crystal lattice filter**

Which of the following filters would be the best choice for use in a 2 meter repeater duplexer?
✓ **A cavity filter**

Which of the following is the common name for a filter network which is equivalent to two L-networks connected back-to-back with the two inductors in series and the capacitors in shunt at the input and output?
✓ **Pi**

Which describes a Pi-L-network used for matching a vacuum tube final amplifier to a 50 ohm unbalanced output?
✓ **A Pi-network with an additional series inductor on the output**

What is one advantage of a Pi-matching network over an L-matching network consisting of a single inductor and a single capacitor?
✓ **The Q of Pi-networks can be varied depending on the component values chosen**

Which mode is most affected by non-linear phase response in a receiver IF filter?
✓ **Digital**

What is a crystal lattice filter?
✓ **A filter with narrow bandwidth and steep skirts made using quartz crystals**

E7D – POWER SUPPLISE AND VOLTAGE REGULATORS; SOLAR ARRAY CHARGE CONTROLLERS

What is one characteristic of a linear electronic voltage regulator?
✓ **The conduction of a control element is varied to maintain a constant output voltage**

What is one characteristic of a switching electronic voltage regulator?
✓ **The controlled device's duty cycle is changed to produce a constant average output voltage**

What device is typically used as a stable reference voltage in a linear voltage regulator?
✓ **A Zener diode**

Which of the following types of linear voltage regulator usually make the most efficient use of the primary power source?

✓ **A series regulator**

Which of the following types of linear voltage regulator places a constant load on the unregulated voltage source?

✓ **A shunt regulator**

What is the purpose of Q1 in the circuit shown in Figure E7-3?

✓ **It increases the current-handling capability of the regulator**

What is the purpose of C2 in the circuit shown in Figure E7-3?

✓ **It bypasses hum around D1**

What type of circuit is shown in Figure E7-3?

✓ **Linear voltage regulator**

What is the main reason to use a charge controller with a solar power system?

✓ **Prevention of battery damage due to overcharge**

What is the primary reason that a high-frequency switching type high voltage power supply can be both less expensive and lighter in weight than a conventional power supply?

✓ **The high frequency inverter design uses much smaller transformers and filter components for an equivalent power output**

What circuit element is controlled by a series analog voltage regulator to maintain a constant output voltage?

✓ **Pass transistor**

What is the drop-out voltage of an analog voltage regulator?

✓ **Minimum input-to-output voltage required to maintain regulation**

What is the equation for calculating power dissipation by a series connected linear voltage regulator?

✓ **Voltage difference from input to output multiplied by output current**

What is one purpose of a "bleeder" resistor in a conventional unregulated power supply?

✓ **To improve output voltage regulation**

What is the purpose of a "step-start" circuit in a high voltage power supply?

✓ **To allow the filter capacitors to charge gradually**

When several electrolytic filter capacitors are connected in series to increase the operating voltage of a power supply filter circuit, why should resistors be connected across each capacitor?

A. To equalize, as much as possible, the voltage drop across each capacitor
B. To provide a safety bleeder to discharge the capacitors when the supply is off
C. To provide a minimum load current to reduce voltage excursions at light loads
✓ **D. All of these choices are correct**

E7E - MODULATION AND DEMODULATION: REACTANCE, PHASE AND BALANCED
MODULATORS; DETECTORS; MIXER STAGES

Which of the following can be used to generate FM phone emissions?
✓ **A reactance modulator on the oscillator**

What is the function of a reactance modulator?
✓ **To produce PM signals by using an electrically variable inductance or capacitance**

How does an analog phase modulator function?
✓ **By varying the tuning of an amplifier tank circuit to produce PM signals**

What is one way a single-sideband phone signal can be generated?
✓ **By using a balanced modulator followed by a filter**

What circuit is added to an FM transmitter to boost the higher audio frequencies?
✓ **A pre-emphasis network**

Why is de-emphasis commonly used in FM communications receivers?
✓ **For compatibility with transmitters using phase modulation**

What is meant by the term baseband in radio communications?
✓ **The frequency components present in the modulating signal**

What are the principal frequencies that appear at the output of a mixer circuit?
✓ **The two input frequencies along with their sum and difference frequencies**

What occurs when an excessive amount of signal energy reaches a mixer circuit?
✓ **Spurious mixer products are generated**

How does a diode detector function?
✓ **By rectification and filtering of RF signals**

Which type of detector is used for demodulating SSB signals?
✓ **Product detector**

What is a frequency discriminator stage in a FM receiver?
✓ **A circuit for detecting FM signals**

What is meant by direct digital conversion as applied to software defined radios?
✓ **Incoming RF is digitized by an analog-to-digital converter without being mixed
with a local oscillator signal**

What kind of digital signal processing audio filter is used to remove unwanted noise
from a received SSB signal?
✓ **An adaptive filter**

What type of digital signal processing filter is used to generate an SSB signal?
✓ **A Hilbert-transform filter**

What is a common method of generating an SSB signal using digital signal processing?
✓ **Combine signals with a quadrature phase relationship**

How frequently must an analog signal be sampled by an analog-to-digital converter so
that the signal can be accurately reproduced?
✓ **At twice the rate of the highest frequency component of the signal**

What is the minimum number of bits required for an analog-to-digital converter to
sample a signal with a range of 1 volt at a resolution of 1 millivolt?
✓ **10 bits**

What function can a Fast Fourier Transform perform?
✓ **Converting digital signals from the time domain to the frequency domain**

What is the function of decimation with regard to digital filters?
✓ **Reducing the effective sample rate by removing samples**

Why is an anti-aliasing digital filter required in a digital decimator?
✓ **It removes high-frequency signal components which would otherwise be
reproduced as lower frequency components**

What aspect of receiver analog-to-digital conversion determines the maximum receive
bandwidth of a Direct Digital Conversion SDR?
✓ **Sample rate**

What sets the minimum detectable signal level for an SDR in the absence of
atmospheric or thermal noise?
✓ **Reference voltage level and sample width in bits**

What digital process is applied to I and Q signals in order to recover the baseband
modulation information?
✓ **Fast Fourier Transform**

What is the function of taps in a digital signal processing filter?

✓ **Provide incremental signal delays for filter algorithms**

Which of the following would allow a digital signal processing filter to create a sharper filter response?

✓ **More taps**

Which of the following is an advantage of a Finite Impulse Response (FIR) filter vs an Infinite Impulse Response (IIR) digital filter?

✓ **FIR filters delay all frequency components of the signal by the same amount**

How might the sampling rate of an existing digital signal be adjusted by a factor of 3/4?

✓ **Interpolate by a factor of three, then decimate by a factor of four**

What do the letters I and Q in I/Q Modulation represent?

✓ **In-phase and Quadrature**

E7G - ACTIVE FILTERS AND OP-AMP CIRCUITS: ACTIVE AUDIO FILTERS; CHARACTERISTICS; BASIC CIRCUIT DESIGN; OPERATIONAL AMPLIFIERS

What is the typical output impedance of an integrated circuit op-amp?

✓ **Very low**

What is the effect of ringing in a filter?

✓ **Undesired oscillations added to the desired signal**

What is the typical input impedance of an integrated circuit op-amp?

✓ **Very high**

What is meant by the term op-amp input offset voltage?

✓ **The differential input voltage needed to bring the open loop output voltage to zero**

How can unwanted ringing and audio instability be prevented in a multi-section op-amp RC audio filter circuit?

✓ **Restrict both gain and Q**

Which of the following is the most appropriate use of an op-amp active filter?

✓ **As an audio filter in a receiver**

What magnitude of voltage gain can be expected from the circuit in Figure E7-4 when R1 is 10 ohms and RF is 470 ohms?

✓ **47**

How does the gain of an ideal operational amplifier vary with frequency?

✓ **It does not vary with frequency**

What will be the output voltage of the circuit shown in Figure E7-4 if R1 is 1000 ohms, RF is 10,000 ohms, and 0.23 volts DC is applied to the input?
✓ **-2.3 volts**

What absolute voltage gain can be expected from the circuit in Figure E7-4 when R1 is 1800 ohms and RF is 68 kilohms?
✓ **38**

What absolute voltage gain can be expected from the circuit in Figure E7-4 when R1 is 3300 ohms and RF is 47 kilohms?
✓ **14**

What is an integrated circuit operational amplifier?
✓ **A high-gain, direct-coupled differential amplifier with very high input impedance and very low output impedance**

E7H - OSCILLATORS AND SIGNAL SOURCES: TYPES OF OSCILLATORS; SYNTHESIZERS AND PHASE-LOCKED LOOPS; DIRECT DIGITAL SYNTHESIZERS; STABILIZING THERMAL DRIFT; MICROPHONICS; HIGH ACCURACY OSCILLATORS

What are three oscillator circuits used in Amateur Radio equipment?
✓ **Colpitts, Hartley and Pierce**

Which describes a microphonic?
✓ **Changes in oscillator frequency due to mechanical vibration**

How is positive feedback supplied in a Hartley oscillator?
✓ **Through a tapped coil**

How is positive feedback supplied in a Colpitts oscillator?
✓ **Through a capacitive divider**

How is positive feedback supplied in a Pierce oscillator?
✓ **Through a quartz crystal**

Which of the following oscillator circuits are commonly used in VFOs?
✓ **Colpitts and Hartley**

How can an oscillator's microphonic responses be reduced?
✓ **Mechanically isolating the oscillator circuitry from its enclosure**

Which of the following components can be used to reduce thermal drift in crystal oscillators?
✓ **NPO capacitors**

What type of frequency synthesizer circuit uses a phase accumulator, lookup table, digital to analog converter, and a low-pass anti-alias filter?
✓ **A direct digital synthesizer**

What information is contained in the lookup table of a direct digital frequency synthesizer?

✓ **The amplitude values that represent a sine-wave output**

What are the major spectral impurity components of direct digital synthesizers?

✓ **Spurious signals at discrete frequencies**

Which of the following must be done to insure that a crystal oscillator provides the frequency specified by the crystal manufacturer?

✓ **Provide the crystal with a specified parallel capacitance**

Which of the following is a technique for providing highly accurate and stable oscillators needed for microwave transmission and reception?

A. Use a GPS signal reference

B. Use a rubidium stabilized reference oscillator

C. Use a temperature-controlled high Q dielectric resonator

✓ **D. All of these choices are correct**

What is a phase-locked loop circuit?

✓ **An electronic servo loop consisting of a phase detector, a low-pass filter, a voltage-controlled oscillator, and a stable reference oscillator**

Which of these functions can be performed by a phase-locked loop?

✓ **Frequency synthesis, FM demodulation**

Subelement E8 - Signals and Emissions [4 Groups]

E8A - AC WAVEFORMS: SINE, SQUARE, SAWTOOTH AND IRREGULAR WAVEFORMS; AC MEASUREMENTS; AVERAGE AND PEP OF RF SIGNALS; FOURIER ANALYSIS; ANALOG TO DIGITAL CONVERSION: DIGITAL TO ANALOG CONVERSION

What is the name of the process that shows that a square wave is made up of a sine wave plus all of its odd harmonics?

✓ **Fourier analysis**

What type of wave has a rise time significantly faster than its fall time (or vice versa)?

✓ **A sawtooth wave**

What type of wave does a Fourier analysis show to be made up of sine waves of a given fundamental frequency plus all of its harmonics?

✓ **A sawtooth wave**

What is "dither" with respect to analog to digital converters?

✓ **A small amount of noise added to the input signal to allow more precise representation of a signal over time**

What would be the most accurate way of measuring the RMS voltage of a complex waveform?

✓ **By measuring the heating effect in a known resistor**

What is the approximate ratio of PEP-to-average power in a typical single-sideband phone signal?

✓ **2.5 to 1**

What determines the PEP-to-average power ratio of a single-sideband phone signal?

✓ **The characteristics of the modulating signal**

Why would a direct or flash conversion analog-to-digital converter be useful for a software defined radio?

✓ **Very high speed allows digitizing high frequencies**

How many levels can an analog-to-digital converter with 8 bit resolution encode?

✓ **256**

What is the purpose of a low pass filter used in conjunction with a digital-to-analog converter?

✓ **Remove harmonics from the output caused by the discrete analog levels generated**

What type of information can be conveyed using digital waveforms?

A. Human speech
B. Video signals
C. Data
✓ **D. All of these choices are correct**

What is an advantage of using digital signals instead of analog signals to convey the same information?

✓ **Digital signals can be regenerated multiple times without error**

Which of these methods is commonly used to convert analog signals to digital signals?

✓ **Sequential sampling**

E8B - MODULATION AND DEMODULATION: MODULATION METHODS; MODULATION INDEX AND DEVIATION RATIO; FREQUENCY AND TIME DIVISION MULTIPLEXING; ORTHOGONAL FREQUENCY DIVISION MULTIPLEXING

What is the term for the ratio between the frequency deviation of an RF carrier wave and the modulating frequency of its corresponding FM-phone signal?

✓ **Modulation index**

How does the modulation index of a phase-modulated emission vary with RF carrier frequency (the modulated frequency)?

✓ **It does not depend on the RF carrier frequency**

What is the modulation index of an FM-phone signal having a maximum frequency deviation of 3000 Hz either side of the carrier frequency when the modulating frequency is 1000 Hz?

✓ **3**

What is the modulation index of an FM-phone signal having a maximum carrier deviation of plus or minus 6 kHz when modulated with a 2 kHz modulating frequency?

✓ **3**

What is the deviation ratio of an FM-phone signal having a maximum frequency swing of plus-or-minus 5 kHz when the maximum modulation frequency is 3 kHz?

✓ **1.67**

What is the deviation ratio of an FM-phone signal having a maximum frequency swing of plus or minus 7.5 kHz when the maximum modulation frequency is 3.5 kHz?

✓ **2.14**

Orthogonal Frequency Division Multiplexing is a technique used for which type of amateur communication?

✓ **High speed digital modes**

What describes Orthogonal Frequency Division Multiplexing?

✓ **A digital modulation technique using subcarriers at frequencies chosen to avoid intersymbol interference**

What is meant by deviation ratio?

✓ **The ratio of the maximum carrier frequency deviation to the highest audio modulating frequency**

What describes frequency division multiplexing?

✓ **Two or more information streams are merged into a baseband, which then modulates the transmitter**

What is digital time division multiplexing?

✓ **Two or more signals are arranged to share discrete time slots of a data transmission**

E8C - DIGITAL SIGNALS: DIGITAL COMMUNICATION MODES; INFORMATION RATE VS BANDWIDTH; ERROR CORRECTION

How is Forward Error Correction implemented?

✓ **By transmitting extra data that may be used to detect and correct transmission errors**

What is the definition of symbol rate in a digital transmission?

✓ **The rate at which the waveform of a transmitted signal changes to convey information**

When performing phase shift keying, why is it advantageous to shift phase precisely at the zero crossing of the RF carrier?

✓ **This results in the least possible transmitted bandwidth for the particular mode**

What technique is used to minimize the bandwidth requirements of a PSK31 signal?

✓ **Use of sinusoidal data pulses**

What is the necessary bandwidth of a 13-WPM international Morse code transmission?

✓ **Approximately 52 Hz**

What is the necessary bandwidth of a 170-hertz shift, 300-baud ASCII transmission?

✓ **0.5 kHz**

What is the necessary bandwidth of a 4800-Hz frequency shift, 9600-baud ASCII FM transmission?

✓ **15.36 kHz**

How does ARQ accomplish error correction?

✓ **If errors are detected, a retransmission is requested**

Which is the name of a digital code where each preceding or following character changes by only one bit?

✓ **Gray code**

What is an advantage of Gray code in digital communications where symbols are transmitted as multiple bits

✓ **It facilitates error detection**

What is the relationship between symbol rate and baud?

✓ **They are the same**

E8D - KEYING DEFECTS AND OVERMODULATION OF DIGITAL SIGNALS; DIGITAL CODES; SPREAD SPECTRUM

Why are received spread spectrum signals resistant to interference?

✓ **Signals not using the spread spectrum algorithm are suppressed in the receiver**

What spread spectrum communications technique uses a high speed binary bit stream to shift the phase of an RF carrier?

✓ **Direct sequence**

How does the spread spectrum technique of frequency hopping work?

✓ **The frequency of the transmitted signal is changed very rapidly according to a particular sequence also used by the receiving station**

What is the primary effect of extremely short rise or fall time on a CW signal?

✓ **The generation of key clicks**

What is the most common method of reducing key clicks?
✓ **Increase keying waveform rise and fall times**

Which of the following indicates likely overmodulation of an AFSK signal such as PSK or MFSK?
✓ **Strong ALC action**

What is a common cause of overmodulation of AFSK signals?
✓ **Excessive transmit audio levels**

What parameter might indicate that excessively high input levels are causing distortion in an AFSK signal?
✓ **Intermodulation Distortion (IMD)**

What is considered a good minimum IMD level for an idling PSK signal?
✓ **-30 dB**

What are some of the differences between the Baudot digital code and ASCII?
✓ **Baudot uses 5 data bits per character, ASCII uses 7 or 8; Baudot uses 2 characters as letters/figures shift codes, ASCII has no letters/figures shift code**

What is one advantage of using ASCII code for data communications?
✓ **It is possible to transmit both upper and lower case text**

What is the advantage of including a parity bit with an ASCII character stream?
✓ **Some types of errors can be detected**

Subelement E9 - Antennas and Transmission Lines [8 Groups]

E9A - BASIC ANTENNA PARAMETERS: RADIATION RESISTANCE, GAIN, BEAMWIDTH, EFFICIENCY, BEAMWIDTH; EFFECTIVE RADIATED POWER, POLARIZATION

What describes an isotropic antenna?
✓ **A theoretical antenna used as a reference for antenna gain**

What antenna has no gain in any direction?
✓ **Isotropic antenna**

Why would one need to know the feed point impedance of an antenna?
✓ **To match impedances in order to minimize standing wave ratio on the transmission line**

Which of the following factors may affect the feed point impedance of an antenna?
✓ **Antenna height, conductor length/diameter ratio and location of nearby conductive objects**

What is included in the total resistance of an antenna system?
✓ **Radiation resistance plus ohmic resistance**

How does the beamwidth of an antenna vary as the gain is increased?
✓ **It decreases**

What is meant by antenna gain?
✓ **The ratio of the radiated signal strength of an antenna in the direction of maximum radiation to that of a reference antenna**

What is meant by antenna bandwidth?
✓ **The frequency range over which an antenna satisfies a performance requirement**

How is antenna efficiency calculated?
✓ **(radiation resistance / total resistance) x 100 per cent**

Which of the following choices is a way to improve the efficiency of a ground-mounted quarter-wave vertical antenna?
✓ **Install a good radial system**

Which of the following factors determines ground losses for a ground-mounted vertical antenna operating in the 3 MHz to 30 MHz range?
✓ **Soil conductivity**

How much gain does an antenna have compared to a 1/2-wavelength dipole when it has 6 dB gain over an isotropic antenna?
✓ **3.85 dB**

How much gain does an antenna have compared to a 1/2-wavelength dipole when it has 12 dB gain over an isotropic antenna?
✓ **9.85 dB**

What is meant by the radiation resistance of an antenna?
✓ **The value of a resistance that would dissipate the same amount of power as that radiated from an antenna**

What is the effective radiated power relative to a dipole of a repeater station with 150 watts transmitter power output, 2 dB feed line loss, 2.2 dB duplexer loss, and 7 dBd antenna gain?
✓ **286 watts**

What is the effective radiated power relative to a dipole of a repeater station with 200 watts transmitter power output, 4 dB feed line loss, 3.2 dB duplexer loss, 0.8 dB circulator loss, and 10 dBd antenna gain?
✓ **317 watts**

What is the effective radiated power of a repeater station with 200 watts transmitter power output, 2 dB feed line loss, 2.8 dB duplexer loss, 1.2 dB circulator loss, and 7 dBi antenna gain?
✓ **252 watts**

What term describes station output, taking into account all gains and losses?
✓ **Effective radiated power**

E9B - ANTENNA PATTERNS: E AND H PLANE PATTERNS; GAIN AS A FUNCTION OF PATTERN; ANTENNA DESIGN

In the antenna radiation pattern shown in Figure E9-1, what is the 3 dB beam-width?
✓ **50 degrees**

In the antenna radiation pattern shown in Figure E9-1, what is the front-to-back ratio?
✓ **18 dB**

In the antenna radiation pattern shown in Figure E9-1, what is the front-to-side ratio?
✓ **14 dB**

What may occur when a directional antenna is operated at different frequencies within the band for which it was designed?
✓ **The gain may change depending on frequency**

What type of antenna pattern over real ground is shown in Figure E9-2?
✓ **Elevation**

What is the elevation angle of peak response in the antenna radiation pattern shown in Figure E9-2?
✓ **7.5 degrees**

How does the total amount of radiation emitted by a directional gain antenna compare with the total amount of radiation emitted from an isotropic antenna, assuming each is driven by the same amount of power?
✓ **They are the same**

How can the approximate beam-width in a given plane of a directional antenna be determined?
✓ **Note the two points where the signal strength of the antenna is 3 dB less than maximum and compute the angular difference**

What type of computer program technique is commonly used for modeling antennas?
✓ **Method of Moments**

What is the principle of a Method of Moments analysis?
✓ **A wire is modeled as a series of segments, each having a uniform value of current**

What is a disadvantage of decreasing the number of wire segments in an antenna model below the guideline of 10 segments per half-wavelength?

✓ **The computed feed point impedance may be incorrect**

What is the far field of an antenna?

✓ **The region where the shape of the antenna pattern is independent of distance**

What does the abbreviation NEC stand for when applied to antenna modeling programs?

✓ **Numerical Electromagnetic Code**

What type of information can be obtained by submitting the details of a proposed new antenna to a modeling program?

A. SWR vs frequency charts

B. Polar plots of the far field elevation and azimuth patterns

C. Antenna gain

✓ **D. All of these choices are correct**

What is the front-to-back ratio of the radiation pattern shown in Figure E9-2?

✓ **28 dB**

How many elevation lobes appear in the forward direction of the antenna radiation pattern shown in Figure E9-2?

✓ **4**

E9C - WIRE AND PHASED ARRAY ANTENNAS: RHOMBIC ANTENNAS; EFFECTS OF GROUND REFLECTIONS; E-OFF ANGLES; PRACTICAL WIRE ANTENNAS: ZEPPS, OCFD, LOOPS

What is the radiation pattern of two 1/4-wavelength vertical antennas spaced 1/2-wavelength apart and fed 180 degrees out of phase?

✓ **A figure-8 oriented along the axis of the array**

What is the radiation pattern of two 1/4 wavelength vertical antennas spaced 1/4 wavelength apart and fed 90 degrees out of phase?

✓ **Cardioid**

What is the radiation pattern of two 1/4 wavelength vertical antennas spaced a 1/2 wavelength apart and fed in phase?

✓ **A Figure-8 broadside to the axis of the array**

What happens to the radiation pattern of an unterminated long wire antenna as the wire length is increased?

✓ **The lobes align more in the direction of the wire**

What is an OCFD antenna?

✓ **A dipole feed approximately 1/3 the way from one end with a 4:1 balun to provide multiband operation**

What is the effect of a terminating resistor on a rhombic antenna?
✓ **It changes the radiation pattern from bidirectional to unidirectional**

What is the approximate feed point impedance at the center of a two-wire folded dipole antenna?
✓ **300 ohms**

What is a folded dipole antenna?
✓ **A dipole consisting of one wavelength of wire forming a very thin loop**

What is a G5RV antenna?
✓ **A multi-band dipole antenna fed with coax and a balun through a selected length of open wire transmission line**

Which of the following describes a Zepp antenna?
✓ **An end fed dipole antenna**

How is the far-field elevation pattern of a vertically polarized antenna affected by being mounted over seawater versus rocky ground?
✓ **The low-angle radiation increases**

Which of the following describes an extended double Zepp antenna?
✓ **A center fed 1.25 wavelength antenna (two 5/8 wave elements in phase)**

What is the main effect of placing a vertical antenna over an imperfect ground?
✓ **It reduces low-angle radiation**

How does the performance of a horizontally polarized antenna mounted on the side of a hill compare with the same antenna mounted on flat ground?
✓ **The main lobe takeoff angle decreases in the downhill direction**

How does the radiation pattern of a horizontally polarized 3-element beam antenna vary with its height above ground?
✓ **The main lobe takeoff angle decreases with increasing height**

E9D - DIRECTIONAL ANTENNAS: GAIN; YAGI ANTENNAS; LOSSES; SWR BANDWIDTH; ANTENNA EFFICIENCY; SHORTENED AND MOBILE ANTENNAS; RF GROUNDING

How does the gain of an ideal parabolic dish antenna change when the operating frequency is doubled?
✓ **Gain increases by 6 dB**

How can linearly polarized Yagi antennas be used to produce circular polarization?
✓ **Arrange two Yagis perpendicular to each other with the driven elements at the same point on the boom fed 90 degrees out of phase**

Where should a high Q loading coil be placed to minimize losses in a shortened vertical antenna?

✓ **Near the center of the vertical radiator**

Why should an HF mobile antenna loading coil have a high ratio of reactance to resistance?

✓ **To minimize losses**

What is a disadvantage of using a multiband trapped antenna?

✓ **It might radiate harmonics**

What happens to the bandwidth of an antenna as it is shortened through the use of loading coils?

✓ **It is decreased**

What is an advantage of using top loading in a shortened HF vertical antenna?

✓ **Improved radiation efficiency**

What happens as the Q of an antenna increases?

✓ **SWR bandwidth decreases**

What is the function of a loading coil used as part of an HF mobile antenna?

✓ **To cancel capacitive reactance**

What happens to feed point impedance at the base of a fixed length HF mobile antenna as the frequency of operation is lowered?

✓ **The radiation resistance decreases and the capacitive reactance increases**

Which of the following types of conductors would be best for minimizing losses in a station's RF ground system?

✓ **A wide flat copper strap**

Which of the following would provide the best RF ground for your station?

✓ **An electrically short connection to 3 or 4 interconnected ground rods driven into the Earth**

What usually occurs if a Yagi antenna is designed solely for maximum forward gain?

✓ **The front-to-back ratio decreases**

E9E - MATCHING: MATCHING ANTENNAS TO FEED LINES; PHASING LINES; POWER DIVIDERS

What system matches a higher impedance transmission line to a lower impedance antenna by connecting the line to the driven element in two places spaced a fraction of a wavelength each side of element center?

✓ **The delta matching system**

What is the name of an antenna matching system that matches an unbalanced feed line to an antenna by feeding the driven element both at the center of the element and at a fraction of a wavelength to one side of center?
✓ **The gamma match**

What is the name of the matching system that uses a section of transmission line connected in parallel with the feed line at or near the feed point?
✓ **The stub match**

What is the purpose of the series capacitor in a gamma-type antenna matching network?
✓ **To cancel the inductive reactance of the matching network**

How must the driven element in a 3-element Yagi be tuned to use a hairpin matching system?
✓ **The driven element reactance must be capacitive**

What is the equivalent lumped-constant network for a hairpin matching system of a 3-element Yagi?
✓ **A shunt inductor**

What term best describes the interactions at the load end of a mismatched transmission line?
✓ **Reflection coefficient**

Which of the following measurements is characteristic of a mismatched transmission line?
✓ **An SWR greater than 1:1**

Which of these matching systems is an effective method of connecting a 50 ohm coaxial cable feed line to a grounded tower so it can be used as a vertical antenna?
✓ **Gamma match**

Which of these choices is an effective way to match an antenna with a 100 ohm feed point impedance to a 50 ohm coaxial cable feed line?
✓ **Insert a 1/4-wavelength piece of 75 ohm coaxial cable transmission line in series between the antenna terminals and the 50 ohm feed cable**

What is an effective way of matching a feed line to a VHF or UHF antenna when the impedances of both the antenna and feed line are unknown?
✓ **Use the universal stub matching technique**

What is the primary purpose of a phasing line when used with an antenna having multiple driven elements?
✓ **It ensures that each driven element operates in concert with the others to create the desired antenna pattern**

What is a use for a Wilkinson divider?

✓ **It is used to divide power equally between two 50 ohm loads while maintaining 50 ohm input impedance**

E9F - TRANSMISSION LINES: CHARACTERISTICS OF OPEN AND SHORTED FEED LINES; 1/8 WAVELENGTH; 1/4 WAVELENGTH; 1/2 WAVELENGTH; FEED LINES: COAX VERSUS OPEN-WIRE; VELOCITY FACTOR; ELECTRICAL LENGTH; COAXIAL CABLE DIELECTRICS; VELOCITY FACTOR

What is the velocity factor of a transmission line?

✓ **The velocity of the wave in the transmission line divided by the velocity of light in a vacuum**

Which of the following determines the velocity factor of a transmission line?

✓ **Dielectric materials used in the line**

Why is the physical length of a coaxial cable transmission line shorter than its electrical length?

✓ **Electrical signals move more slowly in a coaxial cable than in air**

What is the typical velocity factor for a coaxial cable with solid polyethylene dielectric?

✓ **0.66**

What is the approximate physical length of a solid polyethylene dielectric coaxial transmission line that is electrically one-quarter wavelength long at 14.1 MHz?

✓ **3.5 meters**

What is the approximate physical length of an air-insulated, parallel conductor transmission line that is electrically one-half wavelength long at 14.10 MHz?

✓ **10 meters**

How does ladder line compare to small-diameter coaxial cable such as RG-58 at 50 MHz?

✓ **Lower loss**

What is the term for the ratio of the actual speed at which a signal travels through a transmission line to the speed of light in a vacuum?

✓ **Velocity factor**

What is the approximate physical length of a solid polyethylene dielectric coaxial transmission line that is electrically one-quarter wavelength long at 7.2 MHz?

✓ **6.9 meters**

What impedance does a 1/8 wavelength transmission line present to a generator when the line is shorted at the far end?

✓ **An inductive reactance**

What impedance does a 1/8 wavelength transmission line present to a generator when the line is open at the far end?
✓ **A capacitive reactance**

What impedance does a 1/4 wavelength transmission line present to a generator when the line is open at the far end?
✓ **Very low impedance**

What impedance does a 1/4 wavelength transmission line present to a generator when the line is shorted at the far end?
✓ **Very high impedance**

What impedance does a 1/2 wavelength transmission line present to a generator when the line is shorted at the far end?
✓ **Very low impedance**

What impedance does a 1/2 wavelength transmission line present to a generator when the line is open at the far end?
✓ **Very high impedance**

Which of the following is a significant difference between foam dielectric coaxial cable and solid dielectric cable, assuming all other parameters are the same?
A. Foam dielectric has lower safe operating voltage limits
B. Foam dielectric has lower loss per unit of length
C. Foam dielectric has higher velocity factor
✓ **D. All of these choices are correct**

E9G - THE SMITH CHART

Which of the following can be calculated using a Smith chart?
✓ **Impedance along transmission lines**

What type of coordinate system is used in a Smith chart?
✓ **Resistance circles and reactance arcs**

Which of the following is often determined using a Smith chart?
✓ **Impedance and SWR values in transmission lines**

What are the two families of circles and arcs that make up a Smith chart?
✓ **Resistance and reactance**

What type of chart is shown in Figure E9-3?
✓ **Smith chart**

On the Smith chart shown in Figure E9-3, what is the name for the large outer circle on which the reactance arcs terminate?
✓ **Reactance axis**

On the Smith chart shown in Figure E9-3, what is the only straight line shown?
- ✓ **The resistance axis**

What is the process of normalization with regard to a Smith chart?
- ✓ **Reassigning impedance values with regard to the prime center**

What third family of circles is often added to a Smith chart during the process of solving problems?
- ✓ **Standing wave ratio circles**

What do the arcs on a Smith chart represent?
- ✓ **Points with constant reactance**

How are the wavelength scales on a Smith chart calibrated?
- ✓ **In fractions of transmission line electrical wavelength**

E9H - RECEIVING ANTENNAS: RADIO DIRECTION FINDING ANTENNAS; BEVERAGE ANTENNAS; SPECIALIZED RECEIVING ANTENNAS; LONGWIRE RECEIVING ANTENNAS

When constructing a Beverage antenna, which of the following factors should be included in the design to achieve good performance at the desired frequency?
- ✓ **It should be one or more wavelengths long**

Which is generally true for low band (160 meter and 80 meter) receiving antennas?
- ✓ **Atmospheric noise is so high that gain over a dipole is not important**

What is an advantage of using a shielded loop antenna for direction finding?
- ✓ **It is electro statically balanced against ground, giving better nulls**

What is the main drawback of a wire-loop antenna for direction finding?
- ✓ **It has a bidirectional pattern**

What is the triangulation method of direction finding?
- ✓ **Antenna headings from several different receiving locations are used to locate the signal source**

Why is it advisable to use an RF attenuator on a receiver being used for direction finding?
- ✓ **It prevents receiver overload which could make it difficult to determine peaks or nulls**

What is the function of a sense antenna?
- ✓ **It modifies the pattern of a DF antenna array to provide a null in one direction**

Which of the following describes the construction of a receiving loop antenna?
- ✓ **One or more turns of wire wound in the shape of a large open coil**

How can the output voltage of a multiple turn receiving loop antenna be increased?
✓ **By increasing either the number of wire turns in the loop or the area of the loop structure or both**

What characteristic of a cardioid pattern antenna is useful for direction finding?
✓ **A very sharp single null**

Subelement E0 – Safety - [1 Group]

E0A - SAFETY: AMATEUR RADIO SAFETY PRACTICES; RF RADIATION HAZARDS; HAZARDOUS MATERIALS; GROUNDING

What is the primary function of an external earth connection or ground rod?
✓ **Lightning protection**

When evaluating RF exposure levels from your station at a neighbor's home, what must you do?
✓ **Make sure signals from your station are less than the uncontrolled MPE limits**

Which of the following would be a practical way to estimate whether the RF fields produced by an amateur radio station are within permissible MPE limits?
✓ **Use an antenna modeling program to calculate field strength at accessible locations**

When evaluating a site with multiple transmitters operating at the same time, the operators and licensees of which transmitters are responsible for mitigating over-exposure situations?
✓ **Each transmitter that produces 5 percent or more of its MPE limit at accessible locations**

What is one of the potential hazards of using microwaves in the amateur radio bands?
✓ **The high gain antennas commonly used can result in high exposure levels**

Why are there separate electric (E) and magnetic (H) field MPE limits?
A. The body reacts to electromagnetic radiation from both the E and H fields
B. Ground reflections and scattering make the field impedance vary with location
C. E field and H field radiation intensity peaks can occur at different locations
✓ **D. All of these choices are correct**

How may dangerous levels of carbon monoxide from an emergency generator be detected?
✓ **Only with a carbon monoxide detector**

What does SAR measure?
✓ **The rate at which RF energy is absorbed by the body**

Which insulating material commonly used as a thermal conductor for some types of electronic devices is extremely toxic if broken or crushed and the particles are accidentally inhaled?
✓ **Beryllium Oxide**

What toxic material may be present in some electronic components such as high voltage capacitors and transformers?
✓ **Polychlorinated Biphenyls**

Which of the following injuries can result from using high-power UHF or microwave transmitters?
✓ **Localized heating of the body from RF exposure in excess of the MPE limits**

~~~~End of "Questions with Just the Right Answers"~~~~

JUST THE RIGHT ANSWERS WITHOUT THE QUESTIONS

Below are all the multiple choice questions in the Extra Class Question Pool but edited to list only the right answers without questions. One of my readers told me he was stumped on the right answer to one of the Extra Class exam questions until he remembered he had seen it on the "Just the Right Answers" list! He credited this simple trick with helping him pass the exam.

Subelement E1 – Commission's Rules [6 Groups]

E1A - OPERATING STANDARDS: FREQUENCY PRIVILEGES; EMISSION STANDARDS; AUTOMATIC MESSAGE FORWARDING; FREQUENCY SHARING; STATIONS ABOARD SHIPS OR AIRCRAFT

3 kHz below the upper band edge

3 kHz above the lower band edge

No, the sideband will extend beyond the band edge

No, the sideband will extend beyond the edge of the phone band segment

100 watts PEP effective radiated power relative to the gain of a half-wave dipole

At the center frequency of the channel

60 meter band

The control operator of the originating station

Discontinue forwarding the communication as soon as you become aware of it

Its operation must be approved by the master of the ship or the pilot in command of the aircraft

Any FCC-issued amateur license

No, one of the sidebands of the CW signal will be out of the band

Any person holding an FCC issued amateur license or who is authorized for alien reciprocal operation

2.8 kHz

An emission outside its necessary bandwidth that can be reduced or eliminated without affecting the information transmitted

The location is of environmental importance or significant in American history, architecture, or culture

1 mile

An Environmental Assessment must be submitted to the FCC

An area surrounding the National Radio Astronomy Observatory

A. You may have to notify the Federal Aviation Administration and register it with the FCC as required by Part 17 of FCC rules

1.0

The amateur station must avoid transmitting during certain hours on frequencies that cause the interference

Any FCC-licensed amateur station certified by the responsible civil defense organization for the area served

All amateur service frequencies authorized to the control operator

At least 43 dB below

A station controlled indirectly through a control link

The use of devices and procedures for control so that the control operator does not have to be present at a control point

Under automatic control the control operator is not required to be present at the control point

An international amateur radio permit that allows U.S. amateurs to operate in certain countries of the Americas

Never

A control operator must be present at the control point

Direct manipulation of the transmitter by a control operator

3 minutes

29.500 MHz - 29.700 MHz

Only auxiliary, repeater or space stations

CEPT agreement

Communications incidental to the purpose of the amateur service and remarks of a personal nature

You must bring a copy of FCC Public Notice DA 11-221

E1D – AMATEUR SATELLITES; DEFINITIONS AND PURPOSE; LICENSE REQURIEMENTS FOR SPACE STATIONS; AVAILABLE FREQUENCIES AND BANDS; TELECOMMAND AND TELEMETRY OPERATIONS; RESTRICTIONS, AND SPECIAL PROVISIONS; NOTIFICATION REQUIREMENTS

One-way transmission of measurements at a distance from the measuring instrument

A radio communications service using amateur radio stations on satellites

An amateur station that transmits communications to initiate, modify or terminate functions of a space station

An amateur station within 50 km of the Earth's surface intended for communications with amateur stations by means of objects in space

Any class with appropriate operator privileges

The space station must be capable of terminating transmissions by telecommand when directed by the FCC

Only the 40 m, 20 m, 17 m, 15 m, 12 m and 10 m bands

2 meters

70 cm and 13 cm

Any amateur station so designated by the space station licensee, subject to the privileges of the class of operator license held by the control operator

Any amateur station, subject to the privileges of the class of operator license held by the control operator

3

In a question pool maintained by all the VECs

An organization that has entered into an agreement with the FCC to coordinate amateur operator license examinations

The procedure by which a VEC confirms that the VE applicant meets FCC requirements to serve as an examiner

Minimum passing score of 74%

Each administering VE

Immediately terminate the candidate's examination

Relatives of the VE as listed in the FCC rules

Revocation of the VE's amateur station license grant and the suspension of the VE's amateur operator license grant

They must submit the application document to the coordinating VEC according to the coordinating VEC instructions

Three VEs must certify that the examinee is qualified for the license grant and that they have complied with the administering VE requirements

Return the application document to the examinee

Use a real-time video link and the Internet to connect the exam session to the observing VEs

Preparing, processing, administering and coordinating an examination for an amateur radio license

Only on amateur frequencies above 222 MHz

The operating terms and conditions of the Canadian amateur service license, not to exceed U.S. Extra Class privileges

It was purchased in used condition from an amateur operator and is sold to another amateur operator for use at that operator's station

A line roughly parallel to and south of the U.S.-Canadian border

420 MHz - 430 MHz

To provide for experimental amateur communications

When neither the amateur nor his or her employer has a pecuniary interest in the communications

Communications transmitted for hire or material compensation, except as otherwise provided in the rules

A. A station transmitting SS emission must not cause harmful interference to other stations employing other authorized emissions
B. The transmitting station must be in an area regulated by the FCC or in a country that permits SS emissions
C. The transmission must not be used to obscure the meaning of any communication
   ✓ **D. All of these choices are correct**

10 W

It must satisfy the FCC's spurious emission standards when operated at the lesser of 1500 watts or its full output power

Only Technician, General, Advanced or Amateur Extra Class operators

Subelement E2 - Operating Procedures [5 Groups]

E2A - AMATEUR RADIO IN SPACE: AMATEUR SATELLITES; ORBITAL MECHANICS; FREQUENCIES AND MODES; SATELLITE HARDWARE; SATELLITE OPERATIONS; EXPERIMENTAL TELEMETRY APPLICATIONS

From south to north

From north to south

The time it takes for a satellite to complete one revolution around the Earth

The satellite's uplink and downlink frequency bands

The uplink and downlink frequency ranges

A. FM and CW
B. SSB and SSTV
C. PSK and Packet
   ✓ **D. All of these choices are correct**

To avoid reducing the downlink power to all other users

The 23 centimeter and 13 centimeter bands

Because the satellite is spinning

A circularly polarized antenna

By calculations using the Keplerian elements for the specified satellite

Geostationary

APRS

E2B - TELEVISION PRACTICES: FAST SCAN TELEVISION STANDARDS AND TECHNIQUES; SLOW SCAN TELEVISION STANDARDS AND TECHNIQUES

30

525

By scanning odd numbered lines in one field and even numbered lines in the next

Turning off the scanning beam while it is traveling from right to left or from bottom to top

Vestigial sideband reduces bandwidth while allowing for simple video detector circuitry

Amplitude modulation in which one complete sideband and a portion of the other are transmitted

Chroma

A. Frequency-modulated sub-carrier
B. A separate VHF or UHF audio link
C. Frequency modulation of the video carrier
✓ D. All of these choices are correct

No other hardware is needed

3 KHz

To identify the SSTV mode being used

Varying tone frequencies representing the video are transmitted using single sideband

128 or 256

Tone frequency

Specific tone frequencies

NTSC

3 kHz

1255 MHz

They are restricted to phone band segments and their bandwidth can be no greater than that of a voice signal of the same modulation type

E2C – OPERATING METHODS: CONTEST AND DX OPERATING; REMOTE OPERATION TECHNIQUES; CABRILLO FORMAT; QSLING; RF NETWORK CONNECTED SYSTEMS

Operators are permitted to make contacts even if they do not submit a log

The generally prohibited practice of posting one's own call sign and frequency on a spotting network

30 m

Spread spectrum in the 2.4 GHz band

To handle the receiving and sending of confirmation cards for a DX station

In the weak signal segment of the band, with most of the activity near the calling frequency

A standard for submission of electronic contest logs

Contacts between a U.S. station and a non-U.S. station

A standard wireless router running custom software

A. Because the DX station may be transmitting on a frequency that is prohibited to some responding stations
B. To separate the calling stations from the DX station
C. To improve operating efficiency by reducing interference
✓ **D. All of these choices are correct**

Send your full call sign once or twice

Switch to a lower frequency HF band

No additional indicator is required

FSK441

A. 15 second timed transmission sequences with stations alternating based on location
B. Use of high speed CW or digital modes
C. Short transmission with rapidly repeated call signs and signal reports
✓ **D. All of these choices are correct**

JT65

To store digital messages in the satellite for later download by other stations

Store-and-forward

Time synchronous transmissions alternately from each station

AX.25

Unnumbered Information

300 baud packet

An APRS station with a GPS unit can automatically transmit information to show a mobile station's position during the event

Latitude and longitude

It can decode signals many dB below the noise floor using FEC

Multi-tone AFSK

The ability to decode signals which have a very low signal to noise ratio

<u>E2E – OPERATING METHODS: OPERATING HF DIGITAL MODES</u>

FSK

Forward Error Correction

Alternating transmissions at 1 minute intervals

Selective fading has occurred

Winlink

300 baud

316 Hz

PACTOR

PSK31

PSK31

Direct FSK applies the data signal to the transmitter VFO

Automatic

A. Your transmit frequency is incorrect
B. The protocol version you are using is not the supported by the digital station
C. Another station you are unable to hear is using the frequency
✓ **D. All of these choices are correct**

Subelement E3 - Radio Wave Propagation [3 Groups]

<u>E3A - ELECTROMAGNETIC WAVES; EARTH-MOON-EARTH COMMUNICATIONS; METEOR SCATTER; MICROWAVE TROPOSPHERIC AND SCATTER PROPAGATION; AURORA PROPAGATION</u>

12,000 miles, if the Moon is visible by both stations

A fluttery irregular fading

When the Moon is at perigee

Probability of tropospheric propagation

Warm and cold fronts

The rain must be within radio range of both stations

Bodies of water

The E layer

28 MHz - 148 MHz

Temperature inversion

100 miles to 300 miles

The interaction in the E layer of charged particles from the Sun with the Earth's magnetic field

CW

North

A wave consisting of an electric field and a magnetic field oscillating at right angles to each other

Changing electric and magnetic fields propagate the energy

Waves with a rotating electric field

Propagation between two mid-latitude points at approximately the same distance north and south of the magnetic equator

5000 miles

Afternoon or early evening

Independent waves created in the ionosphere that are elliptically polarized

160 meters to 10 meters

20 meters

Receipt of a signal by more than one path

Gray-line

Around the solstices, especially the summer solstice

At twilight and sunrise, D-layer absorption is low while E-layer and F-layer propagation remains high

Any time

Successive ionospheric reflections without an intermediate reflection from the ground

The signal experiences less loss along the path compared to normal skip propagation

They become elliptically polarized

Modeling a radio wave's path through the ionosphere

Increasing disruption of the geomagnetic field

Polar paths

Direction and strength of the interplanetary magnetic field

Southward

By approximately 15 percent of the distance

Class X

An extreme geomagnetic storm

Twice as great

UV emissions at 304 angstroms, correlated to solar flux index

HF propagation

It decreases

Vertical

Downward bending due to density variations in the atmosphere

A solar flare has occurred

Subelement E4 - Amateur Practices [5 Groups]

E4A - TEST EQUIPMENT: ANALOG AND DIGITAL INSTRUMENTS; SPECTRUM AND NETWORK ANALYZERS, ANTENNA ANALYZERS; OSCILLOSCOPES; RF MEASUREMENTS; COMPUTER AIDED MEASUREMENTS

Sampling rate

RF amplitude and frequency

A spectrum analyzer

Analog-to-digital conversion speed of the soundcard

A. Automatic amplitude and frequency numerical readout
B. Storage of traces for future reference
C. Manipulation of time base after trace capture
✓ **D. All of these choices are correct**

False signals are displayed

Antenna analyzers do not need an external RF source

An antenna analyzer

One-half the sample rate

Logic analyzer

Keep the signal ground connection of the probe as short as possible

Attenuate the transmitter output going to the spectrum analyzer

A square wave is displayed and the probe is adjusted until the horizontal portions of the displayed wave are as nearly flat as possible

It divides a higher frequency signal so a low-frequency counter can display the input frequency

It provides improved resolution of low-frequency signals within a comparable time period

E4B – MEASUREMENT TECHNIQUE AND LIMITATIONS: INSTRUMENT ACCURACY AND PERFORMANCE LIMITATIONS; PROBES; TECHNICQUES TO MINIMIZE ERRORS; MEASUREMENT OF "Q"; INSTRUMENT CALIBRATION; S PARAMETERS; VECTOR NETWORK ANALYZERS

Time base accuracy

It is very precise in obtaining a signal null

146.52 Hz

14.652 Hz

1465.20 Hz

75 watts

The port or ports at which measurements are made

High impedance input

There is more power going into the antenna

Modulate the transmitter with two non-harmonically related audio frequencies and observe the RF output with a spectrum analyzer

Connect the antenna feed line directly to the analyzer's connector

The full scale reading of the voltmeter multiplied by its ohms per volt rating will indicate the input impedance of the voltmeter

S21

A less accurate reading results

The bandwidth of the circuit's frequency response

S11

Short circuit, open circuit, and 50 ohms

E4C - RECEIVER PERFORMANCE CHARACTERISTICS, PHASE NOISE, NOISE FLOOR, IMAGE REJECTION, MDS, SIGNAL-TO-NOISE-RATIO; SELECTIVITY; EFFECTS OF SDR RECEIVER NON-LINEARITY

It can cause strong signals on nearby frequencies to interfere with reception of weak signals

A front-end filter or pre-selector

Capture effect

The ratio in dB of the noise generated by the receiver to the theoretical minimum noise

The theoretical noise at the input of a perfect receiver at room temperature

-148 dBm

The minimum discernible signal

The maximum count value of the analog-to-digital converter

Easier for front-end circuitry to eliminate image responses

300 Hz

E4C11 (B)

2.4 kHz

Undesired signals may be heard

It improves dynamic range by attenuating strong signals near the receive frequency

15.210 MHz

Atmospheric noise

Distortion

Analog-to-digital converter sample width in bits

The difference in dB between the noise floor and the level of an incoming signal which will cause 1 dB of gain compression

Cross-modulation of the desired signal and desensitization from strong adjacent signals

When the repeaters are in close proximity and the signals mix in the final amplifier of one or both transmitters

A properly terminated circulator at the output of the transmitter

146.34 MHz and 146.61 MHz

Intermodulation interference

The off-frequency unwanted signal is heard in addition to the desired signal

Nonlinear circuits or devices

To increase rejection of unwanted signals

A pair of 40 dBm signals will theoretically generate a third-order intermodulation product with the same level as the input signals

The third-order product of two signals which are in the band of interest is also likely to be within the band

Desensitization

Strong adjacent channel signals

Decrease the RF bandwidth of the receiver

Ignition noise

A. Broadband white noise
B. Ignition noise
C. Power line noise
✓ **D. All of these choices are correct**

Signals which appear across a wide bandwidth

By connecting the radio's power leads directly to the battery and by installing coaxial capacitors in line with the alternator leads

By installing a brute-force AC-line filter in series with the motor leads

Thunderstorms

By turning off the AC power line main circuit breaker and listening on a battery operated radio

A common-mode signal at the frequency of the radio transmitter

Nearby signals may appear to be excessively wide even if they meet emission standards

A. The interfering signal sounds like AC hum on an AM receiver or a carrier modulated by 60 Hz hum on a SSB or CW receiver
B. The interfering signal may drift slowly across the HF spectrum
C. The interfering signal can be several kHz in width and usually repeats at regular intervals across a HF band
    ✓ **D. All of these choices are correct**

Nearby corroded metal joints are mixing and re-radiating the broadcast signals

A DSP filter can remove the desired signal at the same time as it removes interfering signals

A. Arcing contacts in a thermostatically controlled device
B. A defective doorbell or doorbell transformer inside a nearby residence
C. A malfunctioning illuminated advertising display
✓ **D. All of these choices are correct**

The appearance of unstable modulated or unmodulated signals at specific frequencies

Common mode currents on the shield and conductors

Common-mode current

Subelement E5 - Electrical Principles [4 Groups]

<u>E5A - RESONANCE AND Q: CHARACTERISTICS OF RESONANT CIRCUITS: SERIES AND PARALLEL RESONANCE; DEFINITIONS AND EFFECTS OF Q; HALF-POWER BANDWIDTH; PHASE RELATIONSHIPS IN REACTIVE CIRCUITS</u>

Resonance

The frequency at which the capacitive reactance equals the inductive reactance

Approximately equal to circuit resistance

Approximately equal to circuit resistance

Maximum

It is at a maximum

Minimum

The voltage and current are in phase

Resistance divided by the reactance of either the inductance or capacitance

Reactance of either the inductance or capacitance divided by the resistance

47.3 kHz

31.4 kHz

Internal voltages and circulating currents increase

3.56 MHz

Lower losses

7.12 MHz

Matching bandwidth is decreased

<u>E5B - TIME CONSTANTS AND PHASE RELATIONSHIPS: RLC TIME CONSTANTS; DEFINITION; TIME CONSTANTS IN RL AND RC CIRCUITS; PHASE ANGLE BETWEEN VOLTAGE AND CURRENT; PHASE ANGLES OF SERIES RLC; PHASE ANGLE OF INDUCTANCE VS SUSCEPTANCE; ADMITTANCE AND SUSCEPTANCE</u>

One time constant

One time constant

The sign is reversed

220 seconds

The magnitude of the susceptance is the reciprocal of the magnitude of the reactance

The inverse of reactance

14.0 degrees with the voltage lagging the current

14 degrees with the voltage lagging the current

Current leads voltage by 90 degrees

Voltage leads current by 90 degrees

14 degrees with the voltage leading the current

The inverse of impedance

B

E5C - COORDINATE SYSTEMS AND PHASORS IN ELECTRONICS: RECTANGULAR COORDINATES; POLAR COORDINATES; PHASORS

–jX

By phase angle and amplitude

A positive phase angle

A negative phase angle

Phasor diagram

50 ohms resistance in series with 25 ohms capacitive reactance

A quantity with both magnitude and an angular component

Polar coordinates

Resistive component

Reactive component

The coordinate values along the horizontal and vertical axes

It is equivalent to a pure resistance

Rectangular coordinates

Point 4

Point 3

Point 1

Point 8

<u>E5D - AC AND RF ENERGY IN REAL CIRCUITS: SKIN EFFECT; ELECTROSTATIC AND ELECTROMAGNETIC FIELDS; REACTIVE POWER; POWER FACTOR; ELECTRICAL LENGTH OF CONDUCTORS AT UHF AND MICROWAVE FREQUENCIES</u>

As frequency increases, RF current flows in a thinner layer of the conductor, closer to the surface

To avoid unwanted inductive reactance

Precision printed circuit conductors above a ground plane that provide constant impedance interconnects at microwave frequencies

To reduce phase shift along the connection

Inductance

In a direction determined by the left-hand rule

The amount of current flowing through the conductor

Potential energy

It is repeatedly exchanged between the associated magnetic and electric fields, but is not dissipated

By multiplying the apparent power times the power factor

0.5

80 watts

100 Watts

Wattless, nonproductive power

0.707

0.866

600 watts

355 W

<u>E6A - SEMICONDUCTOR MATERIALS AND DEVICES: SEMICONDUCTOR MATERIALS; GERMANIUM, SILICON, P-TYPE, N-TYPE; TRANSISTOR TYPES: NPN, PNP, JUNCTION, FIELD-EFFECT TRANSISTORS: ENHANCEMENT MODE; DEPLETION MODE; MOS; CMOS; N-CHANNEL; P-CHANNEL</u>

In microwave circuits

N-type

Holes in P-type material and electrons in the N-type material are separated by the applied voltage, widening the depletion region

Acceptor impurity

The change of collector current with respect to emitter current

The change in collector current with respect to base current

Base-to-emitter voltage of approximately 0.6 to 0.7 volts

Alpha cutoff frequency

An FET that exhibits a current flow between source and drain when no gate voltage is applied

4

1

To reduce the chance of the gate insulation being punctured by static discharges or excessive voltages

Complementary Metal-Oxide Semiconductor

An FET has high input impedance; a bipolar transistor has low input impedance

P-type

Free electrons

Gate, drain, source

<u>E6B - DIODES</u>

A constant voltage drop under conditions of varying current

Less forward voltage drop

Tunnel

Varactor diode

A large region of intrinsic material

As a VHF/UHF mixer or detector

Excessive junction temperature

Metal-semiconductor junction

As an RF detector

5

Forward DC bias current

As an RF switch

Forward bias

E6C - DIGITAL ICS: FAMILIES OF DIGITAL ICS; GATES; PROGRAMMABLE LOGIC DEVICES (PLDS)

To prevent input noise from causing unstable output signals

The comparator changes its output state

Logic devices with 0, 1, and high impedance output states

Ability to connect many device outputs to a common bus

Lower power consumption

The input switching threshold is about one-half the power supply voltage

A resistor connected to the positive or negative supply line used to establish a voltage when an input or output is an open circuit

2

A programmable collection of logic gates and circuits in a single integrated circuit

4

5

An integrated circuit logic family using both bipolar and CMOS transistors

It has the high input impedance of CMOS and the low output impedance of bipolar transistors

Complex logic functions can be created in a single integrated circuit

E6D - TOROIDAL AND SOLENOIDAL INDUCTORS: PERMEABILITY, CORE MATERIAL, SELECTING, WINDING; TRANSFORMERS; PIEZOELECTRIC DEVICES

35 turns

Motional capacitance, motional inductance, and loss resistance in series, all in parallel with a shunt capacitor representing electrode and stray capacitance

Mechanical deformation of material by the application of a voltage

Ferrite and brass

Ferrite toroids generally require fewer turns to produce a given inductance value

Permeability

From less than 20 Hz to approximately 300 MHz

Powdered-iron cores generally maintain their characteristics at higher currents

Ferrite beads

Toroidal cores confine most of the magnetic field within the core material

43 turns

The ability of the inductor's core to store magnetic energy has been exceeded

Inter-turn capacitance

Brass

Magnetizing current

Snubber capacitor

Harmonics and distortion could result

E6E - ANALOG ICS: MMICS, CCDS, DEVICE PACKAGES

It samples an analog signal and passes it in stages from the input to the output

DIP

Gallium nitride

50 ohms

2 dB

Controlled gain, low noise figure, and constant input and output impedance over the specified frequency range

Microstrip construction

Through a resistor and/or RF choke connected to the amplifier output lead

Surface mount

Surface mount

A total of two rows of connecting pins placed on opposite sides of the package (Dual In-line Package)

Better dissipation of heat

E6F - OPTICAL COMPONENTS: PHOTOCONDUCTIVE PRINCIPLES AND EFFECTS, PHOTOVOLTAIC SYSTEMS, OPTICAL COUPLERS, OPTICAL SENSORS, AND OPTOISOLATORS; LCDS

The increased conductivity of an illuminated semiconductor

It increases

An LED and a phototransistor

The conversion of light to electrical energy

A device which detects rotation of a control by interrupting a light source with a patterned wheel

A crystalline semiconductor

A device that uses semiconductors to implement the functions of an electromechanical relay

Optoisolators provide a very high degree of electrical isolation between a control circuit and the circuit being switched

The relative fraction of light that is converted to current

Silicon

0.5 V

Electrons

A display utilizing a crystalline liquid and polarizing filters which becomes opaque when voltage is applied

They may be hard view through polarized lenses

Subelement E7 - Practical Circuits [8 Groups]

<u>E7A - DIGITAL CIRCUITS: DIGITAL CIRCUIT PRINCIPLES AND LOGIC CIRCUITS: CLASSES OF LOGIC ELEMENTS; POSITIVE AND NEGATIVE LOGIC; FREQUENCY DIVIDERS; TRUTH TABLES</u>

A flip-flop

It produces one output pulse for every ten input pulses

A flip-flop

2

Astable multivibrator

It switches momentarily to the opposite binary state and then returns to its original state after a set time

It produces logic "0" at its output only when all inputs are logic "1"

It produces logic "1" at its output if any or all inputs are logic "1"

It produces logic "0" at its output if any single input is logic "1"

A list of inputs and corresponding outputs for a digital device

Positive Logic

Positive Logic

<u>E7B - AMPLIFIERS: CLASS OF OPERATION; VACUUM TUBE AND SOLID-STATE CIRCUITS; DISTORTION AND INTERMODULATION; SPURIOUS AND PARASITIC SUPPRESSION; MICROWAVE AMPLIFIERS; SWITCHING-TYPE AMPLIFIERS</u>

More than 180 degrees but less than 360 degrees

A type of amplifier that uses switching technology to achieve high efficiency

A low-pass filter to remove switching signal components

Approximately half-way between saturation and cutoff

Install parasitic suppressors and/or neutralize the stage

Push-pull

Signal distortion and excessive bandwidth

By feeding a 180-degree out-of-phase portion of the output back to the input

The tuning capacitor is adjusted for minimum plate current, and the loading capacitor is adjusted for maximum permissible plate current

Fixed bias

Self bias

Common emitter

Emitter load

The power transistor is at saturation or cut off most of the time, resulting in low power dissipation

Use a resistor in series with the emitter

Transmission of spurious signals

Because they are relatively close in frequency to the desired signal

Low input impedance

E7C - FILTERS AND MATCHING NETWORKS: TYPES OF NETWORKS; TYPES OF FILTERS; FILTER APPLICATIONS; FILTER CHARACTERISTICS; IMPEDANCE MATCHING; DSP FILTERING

A capacitor is connected between the input and ground, another capacitor is connected between the output and ground, and an inductor is connected between input and output

It is a high-pass filter

Greater harmonic suppression

It cancels the reactive part of the impedance and changes the resistive part to a desired value

A Chebyshev filter

Extremely sharp cutoff with one or more notches in the stop band

A notch filter

The relative frequencies of the individual crystals

A variable bandwidth crystal lattice filter

A cavity filter

Pi

A Pi-network with an additional series inductor on the output

The Q of Pi-networks can be varied depending on the component values chosen

Digital

A filter with narrow bandwidth and steep skirts made using quartz crystals

E7D – POWER SUPPLISE AND VOLTAGE REGULATORS; SOLAR ARRAY CHARGE
    CONTROLLERS

The conduction of a control element is varied to maintain a constant output voltage

The controlled device's duty cycle is changed to produce a constant average output voltage

A Zener diode

A series regulator

A shunt regulator

It increases the current-handling capability of the regulator

It bypasses hum around D1

Linear voltage regulator

Prevention of battery damage due to overcharge

The high frequency inverter design uses much smaller transformers and filter components for an equivalent power output

Pass transistor

Minimum input-to-output voltage required to maintain regulation

Voltage difference from input to output multiplied by output current

To improve output voltage regulation

To allow the filter capacitors to charge gradually

A. To equalize, as much as possible, the voltage drop across each capacitor
B. To provide a safety bleeder to discharge the capacitors when the supply is off
C. To provide a minimum load current to reduce voltage excursions at light loads
✓ **D. All of these choices are correct**

E7E - MODULATION AND DEMODULATION: REACTANCE, PHASE AND BALANCED
MODULATORS; DETECTORS; MIXER STAGES

A reactance modulator on the oscillator

To produce PM signals by using an electrically variable inductance or capacitance

By varying the tuning of an amplifier tank circuit to produce PM signals

By using a balanced modulator followed by a filter

A pre-emphasis network

For compatibility with transmitters using phase modulation

The frequency components present in the modulating signal

The two input frequencies along with their sum and difference frequencies

Spurious mixer products are generated

By rectification and filtering of RF signals

Product detector

A circuit for detecting FM signals

E7F – DSP FILTERING AND OTHER OPERATIONS; SOFTWARE DEFINED RADIO
FUNDAMENTALS; DSP MODULATION AND REMODULATION

Incoming RF is digitized by an analog-to-digital converter without being mixed with a local oscillator signal

An adaptive filter

A Hilbert-transform filter

Combine signals with a quadrature phase relationship

At twice the rate of the highest frequency component of the signal

10 bits

Converting digital signals from the time domain to the frequency domain

Reducing the effective sample rate by removing samples

It removes high-frequency signal components which would otherwise be reproduced as lower frequency components

Sample rate

Reference voltage level and sample width in bits

Fast Fourier Transform

Provide incremental signal delays for filter algorithms

More taps

FIR filters delay all frequency components of the signal by the same amount

Interpolate by a factor of three, then decimate by a factor of four

In-phase and Quadrature

E7G - ACTIVE FILTERS AND OP-AMP CIRCUITS: ACTIVE AUDIO FILTERS; CHARACTERISTICS; BASIC CIRCUIT DESIGN; OPERATIONAL AMPLIFIERS

Very low

Undesired oscillations added to the desired signal

Very high

The differential input voltage needed to bring the open loop output voltage to zero

Restrict both gain and Q

As an audio filter in a receiver

47

It does not vary with frequency

-2.3 volts

38

14

A high-gain, direct-coupled differential amplifier with very high input impedance and very low output impedance

Colpitts, Hartley and Pierce

Changes in oscillator frequency due to mechanical vibration

Through a tapped coil

Through a capacitive divider

Through a quartz crystal

Colpitts and Hartley

Mechanically isolating the oscillator circuitry from its enclosure

NPO capacitors

A direct digital synthesizer

The amplitude values that represent a sine-wave output

Spurious signals at discrete frequencies

Provide the crystal with a specified parallel capacitance

A. Use a GPS signal reference
B. Use a rubidium stabilized reference oscillator
C. Use a temperature-controlled high Q dielectric resonator
✓ **D. All of these choices are correct**

An electronic servo loop consisting of a phase detector, a low-pass filter, a voltage-controlled oscillator, and a stable reference oscillator

Frequency synthesis, FM demodulation

Subelement E8 - Signals and Emissions [4 Groups]

E8A - AC WAVEFORMS: SINE, SQUARE, SAWTOOTH AND IRREGULAR WAVEFORMS; AC MEASUREMENTS; AVERAGE AND PEP OF RF SIGNALS; FOURIER ANALYSIS; ANALOG TO DIGITAL CONVERSION: DIGITAL TO ANALOG CONVERSION

Fourier analysis

A sawtooth wave

A sawtooth wave

A small amount of noise added to the input signal to allow more precise representation of a signal over time

By measuring the heating effect in a known resistor

2.5 to 1

The characteristics of the modulating signal

Very high speed allows digitizing high frequencies

256

Remove harmonics from the output caused by the discrete analog levels generated

A. Human speech
B. Video signals
C. Data
✓ D. All of these choices are correct

Digital signals can be regenerated multiple times without error

Sequential sampling

E8B - MODULATION AND DEMODULATION: MODULATION METHODS; MODULATION INDEX AND DEVIATION RATIO; FREQUENCY AND TIME DIVISION MULTIPLEXING; ORTHOGONAL FREQUENCY DIVISION MULTIPLEXING

Modulation index

It does not depend on the RF carrier frequency

3

3

1.67

2.14

High speed digital modes

A digital modulation technique using subcarriers at frequencies chosen to avoid intersymbol interference

The ratio of the maximum carrier frequency deviation to the highest audio modulating frequency

Two or more information streams are merged into a baseband, which then modulates the transmitter

Two or more signals are arranged to share discrete time slots of a data transmission

<u>E8C - DIGITAL SIGNALS: DIGITAL COMMUNICATION MODES; INFORMATION RATE VS BANDWIDTH; ERROR CORRECTION</u>

By transmitting extra data that may be used to detect and correct transmission errors

The rate at which the waveform of a transmitted signal changes to convey information

This results in the least possible transmitted bandwidth for the particular mode

Use of sinusoidal data pulses

Approximately 52 Hz

0.5 kHz

15.36 kHz

If errors are detected, a retransmission is requested

Gray code

It facilitates error detection

They are the same

<u>E8D - KEYING DEFECTS AND OVERMODULATION OF DIGITAL SIGNALS; DIGITAL CODES; SPREAD SPECTRUM</u>

Signals not using the spread spectrum algorithm are suppressed in the receiver

Direct sequence

The frequency of the transmitted signal is changed very rapidly according to a particular sequence also used by the receiving station

The generation of key clicks

Increase keying waveform rise and fall times

Strong ALC action

Excessive transmit audio levels

Intermodulation Distortion (IMD)

-30 dB

Baudot uses 5 data bits per character, ASCII uses 7 or 8; Baudot uses 2 characters as letters/figures shift codes, ASCII has no letters/figures shift code

It is possible to transmit both upper and lower case text

Some types of errors can be detected

Subelement E9 - Antennas and Transmission Lines [8 Groups]

E9A - BASIC ANTENNA PARAMETERS: RADIATION RESISTANCE, GAIN, BEAMWIDTH, EFFICIENCY, BEAMWIDTH; EFFECTIVE RADIATED POWER, POLARIZATION

A theoretical antenna used as a reference for antenna gain

Isotropic antenna

To match impedances in order to minimize standing wave ratio on the transmission line

Antenna height, conductor length/diameter ratio and location of nearby conductive objects

Radiation resistance plus ohmic resistance

It decreases

The ratio of the radiated signal strength of an antenna in the direction of maximum radiation to that of a reference antenna

The frequency range over which an antenna satisfies a performance requirement

(radiation resistance / total resistance) x 100 per cent

Install a good radial system

Soil conductivity

3.85 dB

9.85 dB

The value of a resistance that would dissipate the same amount of power as that radiated from an antenna

286 watts

317 watts

252 watts

Effective radiated power

E9B - ANTENNA PATTERNS: E AND H PLANE PATTERNS; GAIN AS A FUNCTION OF
    PATTERN; ANTENNA DESIGN

50 degrees

18 dB

14 dB

The gain may change depending on frequency

Elevation

7.5 degrees

They are the same

Note the two points where the signal strength of the antenna is 3 dB less than maximum and compute the angular difference

Method of Moments

A wire is modeled as a series of segments, each having a uniform value of current

The computed feed point impedance may be incorrect

The region where the shape of the antenna pattern is independent of distance

Numerical Electromagnetic Code

A. SWR vs frequency charts
B. Polar plots of the far field elevation and azimuth patterns
C. Antenna gain
✓ **D. All of these choices are correct**

28 dB

4

A figure-8 oriented along the axis of the array

Cardioid

A Figure-8 broadside to the axis of the array

The lobes align more in the direction of the wire

A dipole feed approximately 1/3 the way from one end with a 4:1 balun to provide multiband operation

It changes the radiation pattern from bidirectional to unidirectional

300 ohms

A dipole consisting of one wavelength of wire forming a very thin loop

A multi-band dipole antenna fed with coax and a balun through a selected length of open wire transmission line

An end fed dipole antenna

The low-angle radiation increases

A center fed 1.25 wavelength antenna (two 5/8 wave elements in phase)

It reduces low-angle radiation

The main lobe takeoff angle decreases in the downhill direction

The main lobe takeoff angle decreases with increasing height

Gain increases by 6 dB

Arrange two Yagis perpendicular to each other with the driven elements at the same point on the boom fed 90 degrees out of phase

Near the center of the vertical radiator

To minimize losses

It might radiate harmonics

It is decreased

Improved radiation efficiency

SWR bandwidth decreases

To cancel capacitive reactance

The radiation resistance decreases and the capacitive reactance increases

A wide flat copper strap

An electrically short connection to 3 or 4 interconnected ground rods driven into the Earth

The front-to-back ratio decreases

E9E - MATCHING: MATCHING ANTENNAS TO FEED LINES; PHASING LINES; POWER DIVIDERS

The delta matching system

The gamma match

The stub match

To cancel the inductive reactance of the matching network

The driven element reactance must be capacitive

A shunt inductor

Reflection coefficient

An SWR greater than 1:1

Gamma match

Insert a 1/4-wavelength piece of 75 ohm coaxial cable transmission line in series between the antenna terminals and the 50 ohm feed cable

Use the universal stub matching technique

It ensures that each driven element operates in concert with the others to create the desired antenna pattern

It is used to divide power equally between two 50 ohm loads while maintaining 50 ohm input impedance

The velocity of the wave in the transmission line divided by the velocity of light in a vacuum

Dielectric materials used in the line

Electrical signals move more slowly in a coaxial cable than in air

0.66

3.5 meters

10 meters

Lower loss

Velocity factor

6.9 meters

An inductive reactance

A capacitive reactance

Very low impedance

Very high impedance

Very low impedance

Very high impedance

A. Foam dielectric has lower safe operating voltage limits
B. Foam dielectric has lower loss per unit of length
C. Foam dielectric has higher velocity factor
✓ **D. All of these choices are correct**

E9G - THE SMITH CHART

Impedance along transmission lines

Resistance circles and reactance arcs

Impedance and SWR values in transmission lines

Resistance and reactance

Smith chart

Reactance axis

The resistance axis

Reassigning impedance values with regard to the prime center

Standing wave ratio circles

Points with constant reactance

In fractions of transmission line electrical wavelength

E9H - RECEIVING ANTENNAS: RADIO DIRECTION FINDING ANTENNAS; BEVERAGE
ANTENNAS; SPECIALIZED RECEIVING ANTENNAS; LONGWIRE RECEIVING ANTENNAS

It should be one or more wavelengths long

Atmospheric noise is so high that gain over a dipole is not important

It is electro statically balanced against ground, giving better nulls

It has a bidirectional pattern

Antenna headings from several different receiving locations are used to locate the
signal source

It prevents receiver overload which could make it difficult to determine peaks or
nulls

It modifies the pattern of a DF antenna array to provide a null in one direction

One or more turns of wire wound in the shape of a large open coil

By increasing either the number of wire turns in the loop or the area of the loop
structure or both

A very sharp single null

Subelement E0 – Safety - [1 Group]

E0A - SAFETY: AMATEUR RADIO SAFETY PRACTICES; RF RADIATION HAZARDS;
HAZARDOUS MATERIALS; GROUNDING

Lightning protection

Make sure signals from your station are less than the uncontrolled MPE limits

Use an antenna modeling program to calculate field strength at accessible locations

Each transmitter that produces 5 percent or more of its MPE limit at accessible locations

The high gain antennas commonly used can result in high exposure levels

A. The body reacts to electromagnetic radiation from both the E and H fields
B. Ground reflections and scattering make the field impedance vary with location
C. E field and H field radiation intensity peaks can occur at different locations
✓ **D. All of these choices are correct**

Only with a carbon monoxide detector

The rate at which RF energy is absorbed by the body

Beryllium Oxide

Polychlorinated Biphenyls

Localized heating of the body from RF exposure in excess of the MPE limits

~~~~End of "Just the Right Answers without the Questions"~~~~

JUST THE QUESTIONS WITHOUT THE ANSWERS

Here are your flashcards! Below are all the questions but without the multiple choice answers, so you can test your knowledge of the subject matter without benefit of peaking at the multiple choices. Good luck!

Subelement E1 – Commission's Rules [6 Groups]

E1A - OPERATING STANDARDS: FREQUENCY PRIVILEGES; EMISSION STANDARDS; AUTOMATIC MESSAGE FORWARDING; FREQUENCY SHARING; STATIONS ABOARD SHIPS OR AIRCRAFT

When using a transceiver that displays the carrier frequency of phone signals, which of the following displayed frequencies represents the highest frequency at which a properly adjusted USB emission will be totally within the band?

When using a transceiver that displays the carrier frequency of phone signals, which of the following displayed frequencies represents the lowest frequency at which a properly adjusted LSB emission will be totally within the band?

With your transceiver displaying the carrier frequency of phone signals, you hear a station calling CQ on 14.349 MHz USB. Is it legal to return the call using upper sideband on the same frequency?

With your transceiver displaying the carrier frequency of phone signals, you hear a DX station calling CQ on 3.601 MHz LSB. Is it legal to return the call using lower sideband on the same frequency?

What is the maximum power output permitted on the 60 meter band?

Where must the carrier frequency of a CW signal be set to comply with FCC rules for 60 meter operation?

Which amateur band requires transmission on specific channels rather than on a range of frequencies?

If a station in a message forwarding system inadvertently forwards a message that is in violation of FCC rules, who is primarily accountable for the rules violation?

What is the first action you should take if your digital message forwarding station inadvertently forwards a communication that violates FCC rules?

If an amateur station is installed aboard a ship or aircraft, what condition must be met before the station is operated?

Which of the following describes authorization or licensing required when operating an amateur station aboard a U.S.-registered vessel in international waters?

With your transceiver displaying the carrier frequency of CW signals, you hear a DX station's CQ on 3.500 MHz. Is it legal to return the call using CW on the same frequency?

Who must be in physical control of the station apparatus of an amateur station aboard any vessel or craft that is documented or registered in the United States?

What is the maximum bandwidth for a data emission on 60 meters?

E1B – STATION RESTRICTIONS AND SPECIAL OPERATIONS: RESTRICTIONS ON STATION LOCATION; GENERAL OPERATING RESTRICTIONS, SPURIOUS EMISSIONS, CONTROL OPERATOR REIMBURSEMENT; ANTENNA STRUCTURE RESTRICTIONS; RACES OPERATIONS; NATIONAL QUIET ZONE

Which of the following constitutes a spurious emission?

Which of the following factors might cause the physical location of an amateur station apparatus or antenna structure to be restricted?

Within what distance must an amateur station protect an FCC monitoring facility from harmful interference?

What must be done before placing an amateur station within an officially designated wilderness area or wildlife preserve, or an area listed in the National Register of Historical Places?

What is the National Radio Quiet Zone?

Which of the following additional rules apply if you are installing an amateur station antenna at a site at or near a public use airport?

What is the highest modulation index permitted at the highest modulation frequency for angle modulation below 29.0 MHz?

What limitations may the FCC place on an amateur station if its signal causes interference to domestic broadcast reception, assuming that the receivers involved are of good engineering design?

Which amateur stations may be operated under RACES rules?

What frequencies are authorized to an amateur station operating under RACES rules?

What is the permitted mean power of any spurious emission relative to the mean power of the fundamental emission from a station transmitter or external RF amplifier installed after January 1, 2003 and transmitting on a frequency below 30 MHZ?

E1C – DEFINITIONS AND RESTRICTIONS PERTIANING TO LOCAL, AUTOMATIC AND REMOTE CONTROL OPERATION; CONTROL OPERATOR RESPONSIBILITIES FOR REMOTE AND AUTOMATICALLY CONTROLLED STATIONS; IARP AND CEPT LICENSES; THIRD PARTY COMMUNICATIONS OVER AUTOMATICALLY CONTROLLED STATIONS

What is a remotely controlled station?

What is meant by automatic control of a station?

How do the control operator responsibilities of a station under automatic control differ from one under local control?

What is meant by IARP?

When may an automatically controlled station originate third party communications?

Which of the following statements concerning remotely controlled amateur stations is true?

What is meant by local control?

What is the maximum permissible duration of a remotely controlled station's transmissions if its control link malfunctions?

Which of these ranges of frequencies is available for an automatically controlled repeater operating below 30 MHz?

What types of amateur stations may automatically retransmit the radio signals of other amateur stations?

Which of the following operating arrangements allows an FCC-licensed U.S. citizen to operate in many European countries, and alien amateurs from many European countries to operate in the U.S.?

What types of communications may be transmitted to amateur stations in foreign countries?

Which of the following is required in order to operate in accordance with CEPT rules in foreign countries where permitted?

E1D – AMATEUR SATELLITES; DEFINITIONS AND PURPOSE; LICENSE REQURIEMENTS FOR SPACE STATIONS; AVAILABLE FREQUENCIES AND BANDS; TELECOMMAND AND TELEMETRY OPERATIONS; RESTRICTIONS, AND SPECIAL PROVISIONS; NOTIFICATION REQUIREMENTS

What is the definition of the term telemetry?

What is the amateur satellite service?

What is a telecommand station in the amateur satellite service?

What is an Earth station in the amateur satellite service?

What class of licensee is authorized to be the control operator of a space station?

Which of the following is a requirement of a space station?

Which amateur service HF bands have frequencies authorized for space stations?

Which VHF amateur service bands have frequencies available for space stations?

Which UHF amateur service bands have frequencies available for a space station?

Which amateur stations are eligible to be telecommand stations?

Which amateur stations are eligible to operate as Earth stations?

EIE – VOLUNTEER EXAMINER PROGRAM: DEFINITIONS; QUALIFICATIONS; PREPARATION
AND ADMINISTRATION OF EXAMS; ACCREDITATION; QUESTION POOLS;
DOCUMENTATION REQUIREMENTS

What is the minimum number of qualified VEs required to administer an Element 4 amateur operator license examination?

Where are the questions for all written U.S. amateur license examinations listed?

What is a Volunteer Examiner Coordinator?

Which of the following best describes the Volunteer Examiner accreditation process?

What is the minimum passing score on amateur operator license examinations?

Who is responsible for the proper conduct and necessary supervision during an amateur operator license examination session?

What should a VE do if a candidate fails to comply with the examiner's instructions during an amateur operator license examination?

To which of the following examinees may a VE not administer an examination?

What may be the penalty for a VE who fraudulently administers or certifies an examination?

What must the administering VEs do after the administration of a successful examination for an amateur operator license?

What must the VE team do if an examinee scores a passing grade on all examination elements needed for an upgrade or new license?

What must the VE team do with the application form if the examinee does not pass the exam?

Which of these choices is an acceptable method for monitoring the applicants if a VEC opts to conduct an exam session remotely?

For which types of out-of-pocket expenses do the Part 97 rules state that VEs and VECs may be reimbursed?

E1F – MISCELLANEOUS RULES: EXTERNAL RF POWER AMPLIFIERS; BUSINESS COMMUNIATIONS; COMPENSATED COMMUNICATIONS; SPREAD SPECTRUM; AUXILIARY STATIONS; RECIPROCAL OPERATING PRIVILEGES; SPECIAL TEMPORARY AUTHORITY

On what frequencies are spread spectrum transmissions permitted?

What privileges are authorized in the U.S. to persons holding an amateur service license granted by the Government of Canada?

Under what circumstances may a dealer sell an external RF power amplifier capable of operation below 144 MHz if it has not been granted FCC certification?

Which of the following geographic descriptions approximately describes "Line A"?

Amateur stations may not transmit in which of the following frequency segments if they are located in the contiguous 48 states and north of Line A?

Under what circumstances might the FCC issue a Special Temporary Authority (STA) to an amateur station?

When may an amateur station send a message to a business?

Which of the following types of amateur station communications are prohibited?

Which of the following conditions apply when transmitting spread spectrum emission?

What is the maximum permitted transmitter peak envelope power for an amateur station transmitting spread spectrum communications?

Which of the following best describes one of the standards that must be met by an external RF power amplifier if it is to qualify for a grant of FCC certification?

Who may be the control operator of an auxiliary station?

Subelement E2 - Operating Procedures [5 Groups]

E2A - AMATEUR RADIO IN SPACE: AMATEUR SATELLITES; ORBITAL MECHANICS; FREQUENCIES AND MODES; SATELLITE HARDWARE; SATELLITE OPERATIONS; EXPERIMENTAL TELEMETRY APPLICATIONS

What is the direction of an ascending pass for an amateur satellite?

What is the direction of a descending pass for an amateur satellite?

What is the orbital period of an Earth satellite?

What is meant by the term mode as applied to an amateur radio satellite?

What do the letters in a satellite's mode designator specify?

On what band would a satellite receive signals if it were operating in mode U/V?

Why should effective radiated power to a satellite which uses a linear transponder be limited?

What do the terms L band and S band specify with regard to satellite communications?

Why may the received signal from an amateur satellite exhibit a rapidly repeating fading effect?

What type of antenna can be used to minimize the effects of spin modulation and Faraday rotation?

What is one way to predict the location of a satellite at a given time?

What type of satellite appears to stay in one position in the sky?

What technology is used to track, in real time, balloons carrying amateur radio transmitters?

E2B - TELEVISION PRACTICES: FAST SCAN TELEVISION STANDARDS AND TECHNIQUES; SLOW SCAN TELEVISION STANDARDS AND TECHNIQUES

How many times per second is a new frame transmitted in a fast-scan (NTSC) television system?

How many horizontal lines make up a fast-scan (NTSC) television frame?

How is an interlaced scanning pattern generated in a fast-scan (NTSC) television system?

What is blanking in a video signal?

Which of the following is an advantage of using vestigial sideband for standard fast-scan TV transmissions?

What is vestigial sideband modulation?

What is the name of the signal component that carries color information in NTSC video?

Which of the following is a common method of transmitting accompanying audio with amateur fast-scan television?

What hardware, other than a receiver with SSB capability and a suitable computer, is needed to decode SSTV using Digital Radio Mondiale (DRM)?

Which of the following is an acceptable bandwidth for Digital Radio Mondiale (DRM) based voice or SSTV digital transmissions made on the HF amateur bands?

What is the function of the Vertical Interval Signaling (VIS) code sent as part of an SSTV transmission?

How are analog SSTV images typically transmitted on the HF bands?

How many lines are commonly used in each frame of an amateur slow-scan color television picture?

What aspect of an amateur slow-scan television signal encodes the brightness of the picture?

What signals SSTV receiving equipment to begin a new picture line?

Which is a video standard used by North American Fast Scan ATV stations?

What is the approximate bandwidth of a slow-scan TV signal?

On which of the following frequencies is one likely to find FM ATV transmissions?

What special operating frequency restrictions are imposed on slow scan TV transmissions?

E2C – OPERATING METHODS: CONTEST AND DX OPERATING; REMOTE OPERATION TECHNIQUES; CABRILLO FORMAT; QSLING; RF NETWORK CONNECTED SYSTEMS

Which of the following is true about contest operating?

Which of the following best describes the term self-spotting in regards to HF contest operating?

From which of the following bands is amateur radio contesting generally excluded?

What type of transmission is most often used for a ham radio mesh network?

What is the function of a DX QSL Manager?

During a VHF/UHF contest, in which band segment would you expect to find the highest level of activity?

What is the Cabrillo format?

Which of the following contacts may be confirmed through the U.S. QSL bureau system?

What type of equipment is commonly used to implement a ham radio mesh network?

Why might a DX station state that they are listening on another frequency?

How should you generally identify your station when attempting to contact a DX station during a contest or in a pileup?

What might help to restore contact when DX signals become too weak to copy across an entire HF band a few hours after sunset?

What indicator is required to be used by U.S.-licensed operators when operating a station via remote control where the transmitter is located in the U.S.?

E2D - OPERATING METHODS: VHF AND UHF DIGITAL MODES AND PROCEDURES; APRS; EME PROCEDURES, METEOR SCATTER PROCEDURES

Which of the following digital modes is especially designed for use for meteor scatter signals?

Which of the following is a good technique for making meteor scatter contacts?

Which of the following digital modes is especially useful for EME communications?

What is the purpose of digital store-and-forward functions on an Amateur Radio satellite?

Which of the following techniques is normally used by low Earth orbiting digital satellites to relay messages around the world?

Which of the following describes a method of establishing EME contacts?

What digital protocol is used by APRS?

What type of packet frame is used to transmit APRS beacon data?

Which of these digital modes has the fastest data throughput under clear communication conditions?

How can an APRS station be used to help support a public service communications activity?

Which of the following data are used by the APRS network to communicate your location?

How does JT65 improve EME communications?

What type of modulation is used for JT65 contacts?

What is one advantage of using JT65 coding?

E2E – OPERATING METHODS: OPERATING HF DIGITAL MODES

Which type of modulation is common for data emissions below 30 MHz?

What do the letters FEC mean as they relate to digital operation?

How is the timing of JT65 contacts organized?

What is indicated when one of the ellipses in an FSK crossed-ellipse display suddenly disappears?

Which type of digital mode does not support keyboard-to-keyboard operation?

What is the most common data rate used for HF packet?

What is the typical bandwidth of a properly modulated MFSK16 signal?

Which of the following HF digital modes can be used to transfer binary files?

Which of the following HF digital modes uses variable-length coding for bandwidth efficiency?

Which of these digital modes has the narrowest bandwidth?

What is the difference between direct FSK and audio FSK?

Which type of control is used by stations using the Automatic Link Enable (ALE) protocol?

Which of the following is a possible reason that attempts to initiate contact with a digital station on a clear frequency are unsuccessful?

Subelement E3 - Radio Wave Propagation [3 Groups]

E3A - ELECTROMAGNETIC WAVES; EARTH-MOON-EARTH COMMUNICATIONS; METEOR SCATTER; MICROWAVE TROPOSPHERIC AND SCATTER PROPAGATION; AURORA PROPAGATION

What is the approximate maximum separation measured along the surface of the Earth between two stations communicating by Moon bounce?

What characterizes libration fading of an EME signal?

When scheduling EME contacts, which of these conditions will generally result in the least path loss?

What do Hepburn maps predict?

Tropospheric propagation of microwave signals often occurs along what weather related structure?

Which of the following is required for microwave propagation via rain scatter?

Atmospheric ducts capable of propagating microwave signals often form over what geographic feature?

When a meteor strikes the Earth's atmosphere, a cylindrical region of free electrons is formed at what layer of the ionosphere?

Which of the following frequency range is most suited for meteor scatter communications?

Which type of atmospheric structure can create a path for microwave propagation?

What is a typical range for tropospheric propagation of microwave signals?

What is the cause of auroral activity?

Which emission mode is best for aurora propagation?

From the contiguous 48 states, in which approximate direction should an antenna be pointed to take maximum advantage of aurora propagation?

What is an electromagnetic wave?

Which of the following best describes electromagnetic waves traveling in free space?

What is meant by circularly polarized electromagnetic waves?

E3B – TRANSEQUATORIAL PROPAGATION; LONG PATH; GRAY-LINE; MULTI-PATH; ORDINARY AND EXTRAORDINARY WAVES; CHORDAL HOP, SPORADIC E MECHANISMS

What is transequatorial propagation?

What is the approximate maximum range for signals using transequatorial propagation?

What is the best time of day for transequatorial propagation?

What is meant by the terms extraordinary and ordinary waves?

Which amateur bands typically support long-path propagation?

Which of the following amateur bands most frequently provides long-path propagation?

Which of the following could account for hearing an echo on the received signal of a distant station?

What type of HF propagation is probably occurring if radio signals travel along the terminator between daylight and darkness?

At what time of year is Sporadic E propagation most likely to occur?

What is the cause of gray-line propagation?

At twilight and sunrise, D-layer absorption is low while E-layer and F-layer

At what time of day is Sporadic-E propagation most likely to occur?

What is the primary characteristic of chordal hop propagation?

Why is chordal hop propagation desirable?

What happens to linearly polarized radio waves that split into ordinary and extraordinary waves in the ionosphere?

E3C – RADIO-PATH HORIZON; LESS COMMON PROPAGATION MODES; PROPAGATION PREDICTION TECHNIQUES AND MODELING; SPACE WEATHER PARAMETERS AND AMATEUR RADIO

What does the term ray tracing describe in regard to radio communications?

What is indicated by a rising A or K index?

Which of the following signal paths is most likely to experience high levels of absorption when the A index or K index is elevated?

What does the value of Bz (B sub Z) represent?

What orientation of Bz (B sub z) increases the likelihood that incoming particles from the Sun will cause disturbed conditions?

By how much does the VHF/UHF radio horizon distance exceed the geometric horizon?

Which of the following descriptors indicates the greatest solar flare intensity?

What does the space weather term G5 mean?

How does the intensity of an X3 flare compare to that of an X2 flare?

What does the 304A solar parameter measure?

What does VOACAP software model?

How does the maximum distance of ground-wave propagation change when the signal frequency is increased?

What type of polarization is best for ground-wave propagation?

Why does the radio-path horizon distance exceed the geometric horizon?

What might a sudden rise in radio background noise indicate?

Subelement E4 - Amateur Practices [5 Groups]

E4A - TEST EQUIPMENT: ANALOG AND DIGITAL INSTRUMENTS; SPECTRUM AND NETWORK ANALYZERS, ANTENNA ANALYZERS; OSCILLOSCOPES; RF MEASUREMENTS; COMPUTER AIDED MEASUREMENTS

Which of the following parameter determines the bandwidth of a digital or computer-based oscilloscope?

Which of the following parameters would a spectrum analyzer display on the vertical and horizontal axes?

Which of the following test instrument is used to display spurious signals and/or intermodulation distortion products in an SSB transmitter?

What determines the upper frequency limit for a computer soundcard-based oscilloscope program?

What might be an advantage of a digital vs an analog oscilloscope?

What is the effect of aliasing in a digital or computer-based oscilloscope?

Which of the following is an advantage of using an antenna analyzer compared to an SWR bridge to measure antenna SWR?

Which of the following instrument would be best for measuring the SWR of a beam antenna?

When using a computer's soundcard input to digitize signals, what is the highest frequency signal that can be digitized without aliasing?

Which of the following displays multiple digital signal states simultaneously?

Which of the following is good practice when using an oscilloscope probe?

Which of the following procedures is an important precaution to follow when connecting a spectrum analyzer to a transmitter output?

How is the compensation of an oscilloscope probe typically adjusted?

What is the purpose of the prescaler function on a frequency counter?

What is an advantage of a period-measuring frequency counter over a direct-count type?

E4B – MEASUREMENT TECHNIQUE AND LIMITATIONS: INSTRUMENT ACCURACY AND PERFORMANCE LIMITATIONS; PROBES; TECHNICQUES TO MINIMIZE ERRORS; MEASUREMENT OF "Q"; INSTRUMENT CALIBRATION; S PARAMETERS; VECTOR NETWORK ANALYZERS

Which of the following factors most affects the accuracy of a frequency counter?

What is an advantage of using a bridge circuit to measure impedance?

If a frequency counter with a specified accuracy of +/- 1.0 ppm reads 146,520,000 Hz, what is the most the actual frequency being measured could differ from the reading?

If a frequency counter with a specified accuracy of +/- 0.1 ppm reads 146,520,000 Hz, what is the most the actual frequency being measured could differ from the reading?

If a frequency counter with a specified accuracy of +/- 10 ppm reads 146,520,000 Hz, what is the most the actual frequency being measured could differ from the reading?

How much power is being absorbed by the load when a directional power meter connected between a transmitter and a terminating load reads 100 watts forward power and 25 watts reflected power?

What do the subscripts of S parameters represent?

Which of the following is a characteristic of a good DC voltmeter?

What is indicated if the current reading on an RF ammeter placed in series with the antenna feed line of a transmitter increases as the transmitter is tuned to resonance?

Which of the following describes a method to measure intermodulation distortion in an SSB transmitter?

How should an antenna analyzer be connected when measuring antenna resonance and feed point impedance?

What is the significance of voltmeter sensitivity expressed in ohms per volt?

Which S parameter is equivalent to forward gain?

What happens if a dip meter is too tightly coupled to a tuned circuit being checked?

Which of the following can be used as a relative measurement of the Q for a series-tuned circuit?

Which S parameter represents return loss or SWR?

What three test loads are used to calibrate a standard RF vector network analyzer?

E4C - RECEIVER PERFORMANCE CHARACTERISTICS, PHASE NOISE, NOISE FLOOR, IMAGE REJECTION, MDS, SIGNAL-TO-NOISE-RATIO; SELECTIVITY; EFFECTS OF SDR RECEIVER NON-LINEARITY

What is an effect of excessive phase noise in the local oscillator section of a receiver?

Which of the following portions of a receiver can be effective in eliminating image signal interference?

What is the term for the blocking of one FM phone signal by another, stronger FM phone signal?

How is the noise figure of a receiver defined?

What does a value of -174 dBm/Hz represent with regard to the noise floor of a receiver?

A CW receiver with the AGC off has an equivalent input noise power density of -174 dBm/Hz. What would be the level of an unmodulated carrier input to this receiver that would yield an audio output SNR of 0 dB in a 400 Hz noise bandwidth?

What does the MDS of a receiver represent?

An SDR receiver is overloaded when input signals exceed what level?

Which of the following choices is a good reason for selecting a high frequency for the design of the IF in a conventional HF or VHF communications receiver?

Which of the following is a desirable amount of selectivity for an amateur RTTY HF receiver?

Which of the following is a desirable amount of selectivity for an amateur SSB phone receiver?

What is an undesirable effect of using too wide a filter bandwidth in the IF section of a receiver?

How does a narrow-band roofing filter affect receiver performance?

What transmit frequency might generate an image response signal in a receiver tuned to 14.300 MHz and which uses a 455 kHz IF frequency?

What is usually the primary source of noise that is heard from an HF receiver with an antenna connected?

Which of the following is caused by missing codes in an SDR receiver's analog-to-digital converter?

Which of the following has the largest effect on an SDR receiver's linearity?

E4D - RECEIVER PERFORMANCE CHARACTERISTICS: BLOCKING DYNAMIC RANGE; INTERMODULATION AND CROSS-MODULATION INTERFERENCE; 3RD ORDER INTERCEPT; DESENSITIZATION; PRESELECTOR

What is meant by the blocking dynamic range of a receiver?

Which of the following describes two problems caused by poor dynamic range in a communications receiver?

How can intermodulation interference between two repeaters occur?

Which of the following may reduce or eliminate intermodulation interference in a repeater caused by another transmitter operating in close proximity?

What transmitter frequencies would cause an intermodulation-product signal in a receiver tuned to 146.70 MHz when a nearby station transmits on 146.52 MHz?

What is the term for unwanted signals generated by the mixing of two or more signals?

Which describes the most significant effect of an off-frequency signal when it is causing cross-modulation interference to a desired signal?

What causes intermodulation in an electronic circuit?

What is the purpose of the preselector in a communications receiver?

What does a third-order intercept level of 40 dBm mean with respect to receiver performance?

Why are third-order intermodulation products created within a receiver of particular interest compared to other products?

What is the term for the reduction in receiver sensitivity caused by a strong signal near the received frequency?

Which of the following can cause receiver desensitization?

Which of the following is a way to reduce the likelihood of receiver desensitization?

E4E - NOISE SUPPRESSION: SYSTEM NOISE; ELECTRICAL APPLIANCE NOISE; LINE NOISE; LOCATING NOISE SOURCES; DSP NOISE REDUCTION; NOISE BLANKERS; GROUNDING FOR SIGNALS

Which of the following types of receiver noise can often be reduced by use of a receiver noise blanker?

Which of the following types of receiver noise can often be reduced with a DSP noise filter?

Which of the following signals might a receiver noise blanker be able to remove from desired signals?

How can conducted and radiated noise caused by an automobile alternator be suppressed?

How can noise from an electric motor be suppressed?

What is a major cause of atmospheric static?

How can you determine if line noise interference is being generated within your home?

What type of signal is picked up by electrical wiring near a radio antenna?

What undesirable effect can occur when using an IF noise blanker?

What is a common characteristic of interference caused by a touch controlled electrical device?

Which is the most likely cause if you are hearing combinations of local AM broadcast signals within one or more of the MF or HF ham bands?

What is one disadvantage of using some types of automatic DSP notch-filters when attempting to copy CW signals?

What might be the cause of a loud roaring or buzzing AC line interference that comes and goes at intervals?

What is one type of electrical interference that might be caused by the operation of a nearby personal computer?

Which of the following can cause shielded cables to radiate or receive interference?

What current flows equally on all conductors of an unshielded multi-conductor cable?

Subelement E5 - Electrical Principles [4 Groups]

E5A - RESONANCE AND Q: CHARACTERISTICS OF RESONANT CIRCUITS: SERIES AND PARALLEL RESONANCE; DEFINITIONS AND EFFECTS OF Q; HALF-POWER BANDWIDTH; PHASE RELATIONSHIPS IN REACTIVE CIRCUITS

What can cause the voltage across reactances in series to be larger than the voltage applied to them?

What is resonance in an electrical circuit?

What is the magnitude of the impedance of a series RLC circuit at resonance?

What is the magnitude of the impedance of a circuit with a resistor, an inductor and a capacitor all in parallel, at resonance?

What is the magnitude of the current at the input of a series RLC circuit as the frequency goes through resonance?

What is the magnitude of the circulating current within the components of a parallel LC circuit at resonance?

What is the magnitude of the current at the input of a parallel RLC circuit at resonance?

What is the phase relationship between the current through and the voltage across a series resonant circuit at resonance?

How is the Q of an RLC parallel resonant circuit calculated?

How is the Q of an RLC series resonant circuit calculated?

What is the half-power bandwidth of a parallel resonant circuit that has a resonant frequency of 7.1 MHz and a Q of 150?

What is the half-power bandwidth of a parallel resonant circuit that has a resonant frequency of 3.7 MHz and a Q of 118?

What is an effect of increasing Q in a resonant circuit?

What is the resonant frequency of a series RLC circuit if R is 22 ohms, L is 50 microhenrys and C is 40 picofarads?

Which of the following can increase Q for inductors and capacitors?

What is the resonant frequency of a parallel RLC circuit if R is 33 ohms, L is 50 microhenrys and C is 10 picofarads?

What is the result of increasing the Q of an impedance-matching circuit?

E5B - TIME CONSTANTS AND PHASE RELATIONSHIPS: RLC TIME CONSTANTS; DEFINITION; TIME CONSTANTS IN RL AND RC CIRCUITS; PHASE ANGLE BETWEEN VOLTAGE AND CURRENT; PHASE ANGLES OF SERIES RLC; PHASE ANGLE OF INDUCTANCE VS SUSCEPTANCE; ADMITTANCE AND SUSCEPTANCE

What is the term for the time required for the capacitor in an RC circuit to be charged to 63.2% of the applied voltage?

What is the term for the time it takes for a charged capacitor in an RC circuit to discharge to 36.8% of its initial voltage?

What happens to the phase angle of a reactance when it is converted to a susceptance?

What is the time constant of a circuit having two 220 microfarad capacitors and two 1 megohm resistors, all in parallel?

What happens to the magnitude of a reactance when it is converted to a susceptance?

What is susceptance?

What is the phase angle between the voltage across and the current through a series RLC circuit if XC is 500 ohms, R is 1 kilohm, and XL is 250 ohms?

What is the phase angle between the voltage across and the current through a series RLC circuit if XC is 100 ohms, R is 100 ohms, and XL is 75 ohms?

What is the relationship between the current through a capacitor and the voltage across a capacitor?

What is the relationship between the current through an inductor and the voltage across an inductor?

What is the phase angle between the voltage across and the current through a series RLC circuit if XC is 25 ohms, R is 100 ohms, and XL is 50 ohms?

What is admittance?

What letter is commonly used to represent susceptance?

E5C - COORDINATE SYSTEMS AND PHASORS IN ELECTRONICS: RECTANGULAR COORDINATES; POLAR COORDINATES; PHASORS

Which of the following represents a capacitive reactance in rectangular notation?

How are impedances described in polar coordinates?

Which of the following represents an inductive reactance in polar coordinates?

Which of the following represents a capacitive reactance in polar coordinates?

What is the name of the diagram used to show the phase relationship between impedances at a given frequency?

What does the impedance 50–j25 represent?

What is a vector?

What coordinate system is often used to display the phase angle of a circuit containing resistance, inductive and/or capacitive reactance?

When using rectangular coordinates to graph the impedance of a circuit, what does the horizontal axis represent?

When using rectangular coordinates to graph the impedance of a circuit, what does the vertical axis represent?

What do the two numbers that are used to define a point on a graph using rectangular coordinates represent?

If you plot the impedance of a circuit using the rectangular coordinate system and find the impedance point falls on the right side of the graph on the horizontal axis, what do you know about the circuit?

What coordinate system is often used to display the resistive, inductive, and/or capacitive reactance components of impedance?

Which point on Figure E5-2 best represents the impedance of a series circuit consisting of a 400 ohm resistor and a 38 picofarad capacitor at 14 MHz?

Which point in Figure E5-2 best represents the impedance of a series circuit consisting of a 300 ohm resistor and an 18 microhenry inductor at 3.505 MHz?

Which point on Figure E5-2 best represents the impedance of a series circuit consisting of a 300 ohm resistor and a 19 picofarad capacitor at 21.200 MHz?

Which point on Figure E5-2 best represents the impedance of a series circuit consisting of a 300 ohm resistor, a 0.64-microhenry inductor and an 85-picofarad capacitor at 24.900 MHz?

E5D - AC AND RF ENERGY IN REAL CIRCUITS: SKIN EFFECT; ELECTROSTATIC AND ELECTROMAGNETIC FIELDS; REACTIVE POWER; POWER FACTOR; ELECTRICAL LENGTH OF CONDUCTORS AT UHF AND MICROWAVE FREQUENCIES

What is the result of skin effect?

Why is it important to keep lead lengths short for components used in circuits for VHF and above?

What is microstrip?

Why are short connections necessary at microwave frequencies?

Which parasitic characteristic increases with conductor length?

In what direction is the magnetic field oriented about a conductor in relation to the direction of electron flow?

What determines the strength of the magnetic field around a conductor?

What type of energy is stored in an electromagnetic or electrostatic field?

What happens to reactive power in an AC circuit that has both ideal inductors and ideal capacitors?

How can the true power be determined in an AC circuit where the voltage and current are out of phase?

What is the power factor of an R-L circuit having a 60 degree phase angle between the voltage and the current?

How many watts are consumed in a circuit having a power factor of 0.2 if the input is 100-VAC at 4 amperes?

How much power is consumed in a circuit consisting of a 100 ohm resistor in series with a 100 ohm inductive reactance drawing 1 ampere?

What is the power factor of an R-L circuit having a 45 degree phase angle between the voltage and the current?

What is the power factor of an R-L circuit having a 30 degree phase angle between the voltage and the current?

How many watts are consumed in a circuit having a power factor of 0.6 if the input is 200VAC at 5 amperes?

How many watts are consumed in a circuit having a power factor of 0.71 if the apparent power is 500VA?

Subelement E6 - Circuit Components [6 Groups]

E6A - SEMICONDUCTOR MATERIALS AND DEVICES: SEMICONDUCTOR MATERIALS;
    GERMANIUM, SILICON, P-TYPE, N-TYPE; TRANSISTOR TYPES: NPN, PNP, JUNCTION,
    FIELD-EFFECT TRANSISTORS: ENHANCEMENT MODE; DEPLETION MODE; MOS;
    CMOS; N-CHANNEL; P-CHANNEL

In what application is gallium arsenide used as a semiconductor material in preference
to germanium or silicon?

Which of the following semiconductor materials contains excess free electrons?

Why does a PN-junction diode not conduct current when reverse biased?

What is the name given to an impurity atom that adds holes to a semiconductor crystal
structure?

What is the alpha of a bipolar junction transistor?

What is the beta of a bipolar junction transistor?

Which of the following indicates that a silicon NPN junction transistor is biased on?

What term indicates the frequency at which the grounded-base current gain of a
transistor has decreased to 0.7 of the gain obtainable at 1 kHz?

What is a depletion-mode FET?

In Figure E6-2, what is the schematic symbol for an N-channel dual-gate MOSFET?

In Figure E6-2, what is the schematic symbol for a P-channel junction FET?

Why do many MOSFET devices have internally connected Zener diodes on the gates?

What do the initials CMOS stand for?

How does DC input impedance at the gate of a field-effect transistor compare with the
DC input impedance of a bipolar transistor?

Which semiconductor material contains excess holes in the outer shell of electrons?

What are the majority charge carriers in N-type semiconductor material?

What are the names of the three terminals of a field-effect transistor?

E6B - DIODES

What is the most useful characteristic of a Zener diode?

What is an important characteristic of a Schottky diode as compared to an ordinary silicon diode when used as a power supply rectifier?

What special type of diode is capable of both amplification and oscillation?

What type of semiconductor device is designed for use as a voltage-controlled capacitor?

What characteristic of a PIN diode makes it useful as an RF switch or attenuator?

Which of the following is a common use of a hot-carrier diode?

What is the failure mechanism when a junction diode fails due to excessive current?

Which of the following describes a type of semiconductor diode?

What is a common use for point contact diodes?

In Figure E6-3, what is the schematic symbol for a light-emitting diode?

What is used to control the attenuation of RF signals by a PIN diode?

What is one common use for PIN diodes?

What type of bias is required for an LED to emit light?

E6C - DIGITAL ICS: FAMILIES OF DIGITAL ICS; GATES; PROGRAMMABLE LOGIC DEVICES (PLDS)

What is the function of hysteresis in a comparator?

What happens when the level of a comparator's input signal crosses the threshold?

What is tri-state logic?

What is the primary advantage of tri-state logic?

What is an advantage of CMOS logic devices over TTL devices?

Why do CMOS digital integrated circuits have high immunity to noise on the input signal or power supply?

What best describes a pull-up or pull-down resistor?

In Figure E6-5, what is the schematic symbol for a NAND gate?

What is a Programmable Logic Device (PLD)?

In Figure E6-5, what is the schematic symbol for a NOR gate?

In Figure E6-5, what is the schematic symbol for the NOT operation (inverter)?

What is BiCMOS logic?

Which of the following is an advantage of BiCMOS logic?

What is the primary advantage of using a Programmable Gate Array (PGA) in a logic circuit?

E6D - TOROIDAL AND SOLENOIDAL INDUCTORS: PERMEABILITY, CORE MATERIAL, SELECTING, WINDING; TRANSFORMERS; PIEZOELECTRIC DEVICES

How many turns will be required to produce a 5-microhenry inductor using a powdered-iron toroidal core that has an inductance index (A L) value of 40 microhenrys/100 turns?

What is the equivalent circuit of a quartz crystal?

Which of the following is an aspect of the piezoelectric effect?

Which materials are commonly used as a slug core in a variable inductor?

What is one reason for using ferrite cores rather than powdered-iron in an inductor?

What core material property determines the inductance of a toroidal inductor?

What is the usable frequency range of inductors that use toroidal cores, assuming a correct selection of core material for the frequency being used?

What is one reason for using powdered-iron cores rather than ferrite cores in an inductor?

What devices are commonly used as VHF and UHF parasitic suppressors at the input and output terminals of a transistor HF amplifier?

What is a primary advantage of using a toroidal core instead of a solenoidal core in an inductor?

How many turns will be required to produce a 1-mH inductor using a core that has an inductance index (A L) value of 523 millihenrys/1000 turns?

What is the definition of saturation in a ferrite core inductor?

What is the primary cause of inductor self-resonance?

Which type of slug material decreases inductance when inserted into a coil?

What is current in the primary winding of a transformer called if no load is attached to the secondary?

What is the common name for a capacitor connected across a transformer secondary that is used to absorb transient voltage spikes?

Why should core saturation of a conventional impedance matching transformer be avoided?

E6E - ANALOG ICS: MMICS, CCDS, DEVICE PACKAGES

Which of the following is true of a charge-coupled device (CCD)?

Which of the following device packages is a through-hole type?

Which of the following materials is likely to provide the highest frequency of operation when used in MMICs?

Which is the most common input and output impedance of circuits that use MMICs?

Which of the following noise figure values is typical of a low-noise UHF preamplifier?

What characteristics of the MMIC make it a popular choice for VHF through microwave circuits?

Which of the following is typically used to construct a MMIC-based microwave amplifier?

How is voltage from a power supply normally furnished to the most common type of monolithic microwave integrated circuit (MMIC)?

Which of the following component package types would be most suitable for use at frequencies above the HF range?

What is the packaging technique in which leadless components are soldered directly to circuit boards?

What is a characteristic of DIP packaging used for integrated circuits?

Why are high-power RF amplifier ICs and transistors sometimes mounted in ceramic packages?

E6F - OPTICAL COMPONENTS: PHOTOCONDUCTIVE PRINCIPLES AND EFFECTS,
    PHOTOVOLTAIC SYSTEMS, OPTICAL COUPLERS, OPTICAL SENSORS, AND
    OPTOISOLATORS; LCDS

What is photoconductivity?

What happens to the conductivity of a photoconductive material when light shines on it?

What is the most common configuration of an optoisolator or optocoupler?

What is the photovoltaic effect?

Which describes an optical shaft encoder?

Which of these materials is affected the most by photoconductivity?

What is a solid state relay?

Why are optoisolators often used in conjunction with solid state circuits when switching 120VAC?

What is the efficiency of a photovoltaic cell?

What is the most common type of photovoltaic cell used for electrical power generation?

What is the approximate open-circuit voltage produced by a fully-illuminated silicon photovoltaic cell?

What absorbs the energy from light falling on a photovoltaic cell?

What is a liquid crystal display (LCD)?

Which of the following is true of LCD displays?

Subelement E7 - Practical Circuits [8 Groups]

E7A - DIGITAL CIRCUITS: DIGITAL CIRCUIT PRINCIPLES AND LOGIC CIRCUITS: CLASSES OF LOGIC ELEMENTS; POSITIVE AND NEGATIVE LOGIC; FREQUENCY DIVIDERS; TRUTH TABLES

Which is a bi-stable circuit?

What is the function of a decade counter digital IC?

Which of the following can divide the frequency of a pulse train by 2?

How many flip-flops are required to divide a signal frequency by 4?

Which of the following is a circuit that continuously alternates between two states without an external clock?

What is a characteristic of a monostable multivibrator?

What logical operation does a NAND gate perform?

What logical operation does an OR gate perform?

What logical operation is performed by an exclusive NOR gate?

What is a truth table?

What type of logic defines "1" as a high voltage?

What type of logic defines "0" as a high voltage?

<u>E7B - AMPLIFIERS: CLASS OF OPERATION; VACUUM TUBE AND SOLID-STATE CIRCUITS; DISTORTION AND INTERMODULATION; SPURIOUS AND PARASITIC SUPPRESSION; MICROWAVE AMPLIFIERS; SWITCHING-TYPE AMPLIFIERS</u>

For what portion of a signal cycle does a Class AB amplifier operate?

What is a Class D amplifier?

Which of the following components form the output of a class D amplifier circuit?

Where on the load line of a Class A common emitter amplifier would bias normally be set?

What can be done to prevent unwanted oscillations in an RF power amplifier?

Which of the following amplifier types reduces or eliminates even order harmonics?

Which of the following is a likely result when a Class C amplifier is used to amplify a single-sideband phone signal?

How can an RF power amplifier be neutralized?

Which of the following describes how the loading and tuning capacitors are to be adjusted when tuning a vacuum tube RF power amplifier that employs a Pi-network output circuit?

In Figure E7-1, what is the purpose of R1 and R2?

In Figure E7-1, what is the purpose of R3?

What type of amplifier circuit is shown in Figure E7-1?

In Figure E7-2, what is the purpose of R?

Why are switching amplifiers more efficient than linear amplifiers?

What is one way to prevent thermal runaway in a bipolar transistor amplifier?

What is the effect of intermodulation products in a linear power amplifier?

Why are odd-order rather than even-order intermodulation distortion products of concern in linear power amplifiers?

What is a characteristic of a grounded-grid amplifier?

E7C - FILTERS AND MATCHING NETWORKS: TYPES OF NETWORKS; TYPES OF FILTERS; FILTER APPLICATIONS; FILTER CHARACTERISTICS; IMPEDANCE MATCHING; DSP FILTERING

How are the capacitors and inductors of a low-pass filter Pi-network arranged between the network's input and output?

Which of the following is a property of a T-network with series capacitors and a parallel shunt inductor?

What advantage does a Pi-L-network have over a regular Pi-network for impedance matching between the final amplifier of a vacuum-tube transmitter and an antenna?

How does an impedance-matching circuit transform a complex impedance to a resistive impedance?

Which filter type is described as having ripple in the passband and a sharp cutoff?

What are the distinguishing features of an elliptical filter?

What kind of filter would you use to attenuate an interfering carrier signal while receiving an SSB transmission?

Which of the following factors has the greatest effect in helping determine the bandwidth and response shape of a crystal ladder filter?

What is a Jones filter as used as part of an HF receiver IF stage?

Which of the following filters would be the best choice for use in a 2 meter repeater duplexer?

Which of the following is the common name for a filter network which is equivalent to two L-networks connected back-to-back with the two inductors in series and the capacitors in shunt at the input and output?

Which describes a Pi-L-network used for matching a vacuum tube final amplifier to a 50 ohm unbalanced output?

What is one advantage of a Pi-matching network over an L-matching network consisting of a single inductor and a single capacitor?

Which mode is most affected by non-linear phase response in a receiver IF filter?

What is a crystal lattice filter?

E7D – POWER SUPPLISE AND VOLTAGE REGULATORS; SOLAR ARRAY CHARGE CONTROLLERS

What is one characteristic of a linear electronic voltage regulator?

What is one characteristic of a switching electronic voltage regulator?

What device is typically used as a stable reference voltage in a linear voltage regulator?

Which of the following types of linear voltage regulator usually make the most efficient use of the primary power source?

Which of the following types of linear voltage regulator places a constant load on the unregulated voltage source?

What is the purpose of Q1 in the circuit shown in Figure E7-3?

What is the purpose of C2 in the circuit shown in Figure E7-3?

What type of circuit is shown in Figure E7-3?

What is the main reason to use a charge controller with a solar power system?

What is the primary reason that a high-frequency switching type high voltage power supply can be both less expensive and lighter in weight than a conventional power supply?

What circuit element is controlled by a series analog voltage regulator to maintain a constant output voltage?

What is the drop-out voltage of an analog voltage regulator?

What is the equation for calculating power dissipation by a series connected linear voltage regulator?

What is one purpose of a "bleeder" resistor in a conventional unregulated power supply?

What is the purpose of a "step-start" circuit in a high voltage power supply?

When several electrolytic filter capacitors are connected in series to increase the operating voltage of a power supply filter circuit, why should resistors be connected across each capacitor?

E7E - MODULATION AND DEMODULATION: REACTANCE, PHASE AND BALANCED
    MODULATORS; DETECTORS; MIXER STAGES

Which of the following can be used to generate FM phone emissions?

What is the function of a reactance modulator?

How does an analog phase modulator function?

What is one way a single-sideband phone signal can be generated?

What circuit is added to an FM transmitter to boost the higher audio frequencies?

Why is de-emphasis commonly used in FM communications receivers?

What is meant by the term baseband in radio communications?

What are the principal frequencies that appear at the output of a mixer circuit?

What occurs when an excessive amount of signal energy reaches a mixer circuit?

How does a diode detector function?

Which type of detector is used for demodulating SSB signals?

What is a frequency discriminator stage in a FM receiver?

E7F – DSP FILTERING AND OTHER OPERATIONS; SOFTWARE DEFINED RADIO FUNDAMENTALS; DSP MODULATION AND REMODULATION

What is meant by direct digital conversion as applied to software defined radios?

What kind of digital signal processing audio filter is used to remove unwanted noise from a received SSB signal?

What type of digital signal processing filter is used to generate an SSB signal?

What is a common method of generating an SSB signal using digital signal processing?

How frequently must an analog signal be sampled by an analog-to-digital converter so that the signal can be accurately reproduced?

What is the minimum number of bits required for an analog-to-digital converter to sample a signal with a range of 1 volt at a resolution of 1 millivolt?

What function can a Fast Fourier Transform perform?

What is the function of decimation with regard to digital filters?

Why is an anti-aliasing digital filter required in a digital decimator?

What aspect of receiver analog-to-digital conversion determines the maximum receive bandwidth of a Direct Digital Conversion SDR?

What sets the minimum detectable signal level for an SDR in the absence of atmospheric or thermal noise?

What digital process is applied to I and Q signals in order to recover the baseband modulation information?

What is the function of taps in a digital signal processing filter?

Which of the following would allow a digital signal processing filter to create a sharper filter response?

Which of the following is an advantage of a Finite Impulse Response (FIR) filter vs an Infinite Impulse Response (IIR) digital filter?

How might the sampling rate of an existing digital signal be adjusted by a factor of 3/4?

What do the letters I and Q in I/Q Modulation represent?

E7G - ACTIVE FILTERS AND OP-AMP CIRCUITS: ACTIVE AUDIO FILTERS; CHARACTERISTICS; BASIC CIRCUIT DESIGN; OPERATIONAL AMPLIFIERS

What is the typical output impedance of an integrated circuit op-amp?

What is the effect of ringing in a filter?

What is the typical input impedance of an integrated circuit op-amp?

What is meant by the term op-amp input offset voltage?

How can unwanted ringing and audio instability be prevented in a multi-section op-amp RC audio filter circuit?

Which of the following is the most appropriate use of an op-amp active filter?

What magnitude of voltage gain can be expected from the circuit in Figure E7-4 when R1 is 10 ohms and RF is 470 ohms?

How does the gain of an ideal operational amplifier vary with frequency?

What will be the output voltage of the circuit shown in Figure E7-4 if R1 is 1000 ohms, RF is 10,000 ohms, and 0.23 volts DC is applied to the input?

E7G10 (C)

What absolute voltage gain can be expected from the circuit in Figure E7-4 when R1 is 1800 ohms and RF is 68 kilohms?

What absolute voltage gain can be expected from the circuit in Figure E7-4 when R1 is 3300 ohms and RF is 47 kilohms?

What is an integrated circuit operational amplifier?

E7H - OSCILLATORS AND SIGNAL SOURCES: TYPES OF OSCILLATORS; SYNTHESIZERS AND PHASE-LOCKED LOOPS; DIRECT DIGITAL SYNTHESIZERS; STABILIZING THERMAL DRIFT; MICROPHONICS; HIGH ACCURACY OSCILLATORS

What are three oscillator circuits used in Amateur Radio equipment?

Which describes a microphonic?

How is positive feedback supplied in a Hartley oscillator?

How is positive feedback supplied in a Colpitts oscillator?

How is positive feedback supplied in a Pierce oscillator?

Which of the following oscillator circuits are commonly used in VFOs?

How can an oscillator's microphonic responses be reduced?

Which of the following components can be used to reduce thermal drift in crystal oscillators?

What type of frequency synthesizer circuit uses a phase accumulator, lookup table, digital to analog converter, and a low-pass anti-alias filter?

What information is contained in the lookup table of a direct digital frequency synthesizer?

What are the major spectral impurity components of direct digital synthesizers?

Which of the following must be done to insure that a crystal oscillator provides the frequency specified by the crystal manufacturer?

Which of the following is a technique for providing highly accurate and stable oscillators needed for microwave transmission and reception?

What is a phase-locked loop circuit?

Which of these functions can be performed by a phase-locked loop?

Subelement E8 - Signals and Emissions [4 Groups]

E8A - AC WAVEFORMS: SINE, SQUARE, SAWTOOTH AND IRREGULAR WAVEFORMS; AC MEASUREMENTS; AVERAGE AND PEP OF RF SIGNALS; FOURIER ANALYSIS; ANALOG TO DIGITAL CONVERSION: DIGITAL TO ANALOG CONVERSION

What is the name of the process that shows that a square wave is made up of a sine wave plus all of its odd harmonics?

What type of wave has a rise time significantly faster than its fall time (or vice versa)?

What type of wave does a Fourier analysis show to be made up of sine waves of a given fundamental frequency plus all of its harmonics?

What is "dither" with respect to analog to digital converters?

What would be the most accurate way of measuring the RMS voltage of a complex waveform?

What is the approximate ratio of PEP-to-average power in a typical single-sideband phone signal?

What determines the PEP-to-average power ratio of a single-sideband phone signal?

Why would a direct or flash conversion analog-to-digital converter be useful for a software defined radio?

How many levels can an analog-to-digital converter with 8 bit resolution encode?

What is the purpose of a low pass filter used in conjunction with a digital-to-analog converter?

What type of information can be conveyed using digital waveforms?

What is an advantage of using digital signals instead of analog signals to convey the same information?

Which of these methods is commonly used to convert analog signals to digital signals?

E8B - MODULATION AND DEMODULATION: MODULATION METHODS; MODULATION INDEX AND DEVIATION RATIO; FREQUENCY AND TIME DIVISION MULTIPLEXING; ORTHOGONAL FREQUENCY DIVISION MULTIPLEXING

What is the term for the ratio between the frequency deviation of an RF carrier wave and the modulating frequency of its corresponding FM-phone signal?

How does the modulation index of a phase-modulated emission vary with RF carrier frequency (the modulated frequency)?

What is the modulation index of an FM-phone signal having a maximum frequency deviation of 3000 Hz either side of the carrier frequency when the modulating frequency is 1000 Hz?

What is the modulation index of an FM-phone signal having a maximum carrier deviation of plus or minus 6 kHz when modulated with a 2 kHz modulating frequency?

What is the deviation ratio of an FM-phone signal having a maximum frequency swing of plus-or-minus 5 kHz when the maximum modulation frequency is 3 kHz?

What is the deviation ratio of an FM-phone signal having a maximum frequency swing of plus or minus 7.5 kHz when the maximum modulation frequency is 3.5 kHz?

Orthogonal Frequency Division Multiplexing is a technique used for which type of amateur communication?

What describes Orthogonal Frequency Division Multiplexing?

What is meant by deviation ratio?

What describes frequency division multiplexing?

What is digital time division multiplexing?

E8C - DIGITAL SIGNALS: DIGITAL COMMUNICATION MODES; INFORMATION RATE VS
   BANDWIDTH; ERROR CORRECTION

How is Forward Error Correction implemented?

What is the definition of symbol rate in a digital transmission?

When performing phase shift keying, why is it advantageous to shift phase precisely at
the zero crossing of the RF carrier?

What technique is used to minimize the bandwidth requirements of a PSK31 signal?

What is the necessary bandwidth of a 13-WPM international Morse code transmission?

What is the necessary bandwidth of a 170-hertz shift, 300-baud ASCII transmission?

What is the necessary bandwidth of a 4800-Hz frequency shift, 9600-baud ASCII FM
transmission?

How does ARQ accomplish error correction?

Which is the name of a digital code where each preceding or following character
changes by only one bit?

What is an advantage of Gray code in digital communications where symbols are
transmitted as multiple bits

What is the relationship between symbol rate and baud?

E8D - KEYING DEFECTS AND OVERMODULATION OF DIGITAL SIGNALS; DIGITAL CODES;
   SPREAD SPECTRUM

Why are received spread spectrum signals resistant to interference?

What spread spectrum communications technique uses a high speed binary bit stream
to shift the phase of an RF carrier?

How does the spread spectrum technique of frequency hopping work?

What is the primary effect of extremely short rise or fall time on a CW signal?

What is the most common method of reducing key clicks?

Which of the following indicates likely overmodulation of an AFSK signal such as PSK or MFSK?

What is a common cause of overmodulation of AFSK signals?

What parameter might indicate that excessively high input levels are causing distortion in an AFSK signal?

What is considered a good minimum IMD level for an idling PSK signal?

What are some of the differences between the Baudot digital code and ASCII?

What is one advantage of using ASCII code for data communications?

What is the advantage of including a parity bit with an ASCII character stream?

Subelement E9 - Antennas and Transmission Lines [8 Groups]

E9A - BASIC ANTENNA PARAMETERS: RADIATION RESISTANCE, GAIN, BEAMWIDTH, EFFICIENCY, BEAMWIDTH; EFFECTIVE RADIATED POWER, POLARIZATION

What describes an isotropic antenna?

What antenna has no gain in any direction?

Why would one need to know the feed point impedance of an antenna?

Which of the following factors may affect the feed point impedance of an antenna?

What is included in the total resistance of an antenna system?

How does the beamwidth of an antenna vary as the gain is increased?

What is meant by antenna gain?

What is meant by antenna bandwidth?

How is antenna efficiency calculated?

Which of the following choices is a way to improve the efficiency of a ground-mounted quarter-wave vertical antenna?

Which of the following factors determines ground losses for a ground-mounted vertical antenna operating in the 3 MHz to 30 MHz range?

How much gain does an antenna have compared to a 1/2-wavelength dipole when it has 6 dB gain over an isotropic antenna?

How much gain does an antenna have compared to a 1/2-wavelength dipole when it has 12 dB gain over an isotropic antenna?

What is meant by the radiation resistance of an antenna?

What is the effective radiated power relative to a dipole of a repeater station with 150 watts transmitter power output, 2 dB feed line loss, 2.2 dB duplexer loss, and 7 dBd antenna gain?

What is the effective radiated power relative to a dipole of a repeater station with 200 watts transmitter power output, 4 dB feed line loss, 3.2 dB duplexer loss, 0.8 dB circulator loss, and 10 dBd antenna gain?

What is the effective radiated power of a repeater station with 200 watts transmitter power output, 2 dB feed line loss, 2.8 dB duplexer loss, 1.2 dB circulator loss, and 7 dBi antenna gain?

What term describes station output, taking into account all gains and losses?

E9B - ANTENNA PATTERNS: E AND H PLANE PATTERNS; GAIN AS A FUNCTION OF PATTERN; ANTENNA DESIGN

In the antenna radiation pattern shown in Figure E9-1, what is the 3 dB beam-width?

In the antenna radiation pattern shown in Figure E9-1, what is the front-to-back ratio?

In the antenna radiation pattern shown in Figure E9-1, what is the front-to-side ratio?

What may occur when a directional antenna is operated at different frequencies within the band for which it was designed?

What type of antenna pattern over real ground is shown in Figure E9-2?

What is the elevation angle of peak response in the antenna radiation pattern shown in Figure E9-2?

How does the total amount of radiation emitted by a directional gain antenna compare with the total amount of radiation emitted from an isotropic antenna, assuming each is driven by the same amount of power?

How can the approximate beam-width in a given plane of a directional antenna be determined?

What type of computer program technique is commonly used for modeling antennas?

What is the principle of a Method of Moments analysis?

What is a disadvantage of decreasing the number of wire segments in an antenna model below the guideline of 10 segments per half-wavelength?

What is the far field of an antenna?

What does the abbreviation NEC stand for when applied to antenna modeling programs?

What type of information can be obtained by submitting the details of a proposed new antenna to a modeling program?

What is the front-to-back ratio of the radiation pattern shown in Figure E9-2?

How many elevation lobes appear in the forward direction of the antenna radiation pattern shown in Figure E9-2?

E9C - WIRE AND PHASED ARRAY ANTENNAS: RHOMBIC ANTENNAS; EFFECTS OF
GROUND REFLECTIONS; E-OFF ANGLES; PRACTICAL WIRE ANTENNAS: ZEPPS, OCFD,
LOOPS

What is the radiation pattern of two 1/4-wavelength vertical antennas spaced 1/2-wavelength apart and fed 180 degrees out of phase?

What is the radiation pattern of two 1/4 wavelength vertical antennas spaced 1/4 wavelength apart and fed 90 degrees out of phase?

What is the radiation pattern of two 1/4 wavelength vertical antennas spaced a 1/2 wavelength apart and fed in phase?

What happens to the radiation pattern of an unterminated long wire antenna as the wire length is increased?

What is an OCFD antenna?

What is the effect of a terminating resistor on a rhombic antenna?

What is the approximate feed point impedance at the center of a two-wire folded dipole antenna?

What is a folded dipole antenna?

What is a G5RV antenna?

Which of the following describes a Zepp antenna?

How is the far-field elevation pattern of a vertically polarized antenna affected by being mounted over seawater versus rocky ground?

Which of the following describes an extended double Zepp antenna?

What is the main effect of placing a vertical antenna over an imperfect ground?

How does the performance of a horizontally polarized antenna mounted on the side of a hill compare with the same antenna mounted on flat ground?

How does the radiation pattern of a horizontally polarized 3-element beam antenna vary with its height above ground?

E9D - DIRECTIONAL ANTENNAS: GAIN; YAGI ANTENNAS; LOSSES; SWR BANDWIDTH; ANTENNA EFFICIENCY; SHORTENED AND MOBILE ANTENNAS; RF GROUNDING

How does the gain of an ideal parabolic dish antenna change when the operating frequency is doubled?

How can linearly polarized Yagi antennas be used to produce circular polarization?

Where should a high Q loading coil be placed to minimize losses in a shortened vertical antenna?

Why should an HF mobile antenna loading coil have a high ratio of reactance to resistance?

What is a disadvantage of using a multiband trapped antenna?

What happens to the bandwidth of an antenna as it is shortened through the use of loading coils?

What is an advantage of using top loading in a shortened HF vertical antenna?

What happens as the Q of an antenna increases?

What is the function of a loading coil used as part of an HF mobile antenna?

What happens to feed point impedance at the base of a fixed length HF mobile antenna as the frequency of operation is lowered?

Which of the following types of conductors would be best for minimizing losses in a station's RF ground system?

Which of the following would provide the best RF ground for your station?

What usually occurs if a Yagi antenna is designed solely for maximum forward gain?

E9E - MATCHING: MATCHING ANTENNAS TO FEED LINES; PHASING LINES; POWER DIVIDERS

What system matches a higher impedance transmission line to a lower impedance antenna by connecting the line to the driven element in two places spaced a fraction of a wavelength each side of element center?

What is the name of an antenna matching system that matches an unbalanced feed line to an antenna by feeding the driven element both at the center of the element and at a fraction of a wavelength to one side of center?

What is the name of the matching system that uses a section of transmission line connected in parallel with the feed line at or near the feed point?

What is the purpose of the series capacitor in a gamma-type antenna matching network?

How must the driven element in a 3-element Yagi be tuned to use a hairpin matching system?

What is the equivalent lumped-constant network for a hairpin matching system of a 3-element Yagi?

What term best describes the interactions at the load end of a mismatched transmission line?

Which of the following measurements is characteristic of a mismatched transmission line?

Which of these matching systems is an effective method of connecting a 50 ohm coaxial cable feed line to a grounded tower so it can be used as a vertical antenna?

Which of these choices is an effective way to match an antenna with a 100 ohm feed point impedance to a 50 ohm coaxial cable feed line?

What is an effective way of matching a feed line to a VHF or UHF antenna when the impedances of both the antenna and feed line are unknown?

What is the primary purpose of a phasing line when used with an antenna having multiple driven elements?

What is a use for a Wilkinson divider?

E9F - TRANSMISSION LINES: CHARACTERISTICS OF OPEN AND SHORTED FEED LINES; 1/8 WAVELENGTH; 1/4 WAVELENGTH; 1/2 WAVELENGTH; FEED LINES: COAX VERSUS OPEN-WIRE; VELOCITY FACTOR; ELECTRICAL LENGTH; COAXIAL CABLE DIELECTRICS; VELOCITY FACTOR

What is the velocity factor of a transmission line?

Which of the following determines the velocity factor of a transmission line?

Why is the physical length of a coaxial cable transmission line shorter than its electrical length?

What is the typical velocity factor for a coaxial cable with solid polyethylene dielectric?

What is the approximate physical length of a solid polyethylene dielectric coaxial transmission line that is electrically one-quarter wavelength long at 14.1 MHz?

What is the approximate physical length of an air-insulated, parallel conductor transmission line that is electrically one-half wavelength long at 14.10 MHz?

How does ladder line compare to small-diameter coaxial cable such as RG-58 at 50 MHz?

What is the term for the ratio of the actual speed at which a signal travels through a transmission line to the speed of light in a vacuum?

What is the approximate physical length of a solid polyethylene dielectric coaxial transmission line that is electrically one-quarter wavelength long at 7.2 MHz?

What impedance does a 1/8 wavelength transmission line present to a generator when the line is shorted at the far end?

What impedance does a 1/8 wavelength transmission line present to a generator when the line is open at the far end?

What impedance does a 1/4 wavelength transmission line present to a generator when the line is open at the far end?

What impedance does a 1/4 wavelength transmission line present to a generator when the line is shorted at the far end?

What impedance does a 1/2 wavelength transmission line present to a generator when the line is shorted at the far end?

What impedance does a 1/2 wavelength transmission line present to a generator when the line is open at the far end?

Which of the following is a significant difference between foam dielectric coaxial cable and solid dielectric cable, assuming all other parameters are the same?

E9G - THE SMITH CHART

Which of the following can be calculated using a Smith chart?

What type of coordinate system is used in a Smith chart?

Which of the following is often determined using a Smith chart?

What are the two families of circles and arcs that make up a Smith chart?

What type of chart is shown in Figure E9-3?

On the Smith chart shown in Figure E9-3, what is the name for the large outer circle on which the reactance arcs terminate?

On the Smith chart shown in Figure E9-3, what is the only straight line shown?

What is the process of normalization with regard to a Smith chart?

What third family of circles is often added to a Smith chart during the process of solving problems?

What do the arcs on a Smith chart represent?

How are the wavelength scales on a Smith chart calibrated?

E9H - RECEIVING ANTENNAS: RADIO DIRECTION FINDING ANTENNAS; BEVERAGE ANTENNAS; SPECIALIZED RECEIVING ANTENNAS; LONGWIRE RECEIVING ANTENNAS

When constructing a Beverage antenna, which of the following factors should be included in the design to achieve good performance at the desired frequency?

Which is generally true for low band (160 meter and 80 meter) receiving antennas?

What is an advantage of using a shielded loop antenna for direction finding?

What is the main drawback of a wire-loop antenna for direction finding?

What is the triangulation method of direction finding?

Why is it advisable to use an RF attenuator on a receiver being used for direction finding?

What is the function of a sense antenna?

Which of the following describes the construction of a receiving loop antenna?

How can the output voltage of a multiple turn receiving loop antenna be increased?

What characteristic of a cardioid pattern antenna is useful for direction finding?

Subelement E0 – Safety - [1 Group]

E0A - SAFETY: AMATEUR RADIO SAFETY PRACTICES; RF RADIATION HAZARDS; HAZARDOUS MATERIALS; GROUNDING

What is the primary function of an external earth connection or ground rod?

When evaluating RF exposure levels from your station at a neighbor's home, what must you do?

Which of the following would be a practical way to estimate whether the RF fields produced by an amateur radio station are within permissible MPE limits?

When evaluating a site with multiple transmitters operating at the same time, the operators and licensees of which transmitters are responsible for mitigating over-exposure situations?

What is one of the potential hazards of using microwaves in the amateur radio bands?

Why are there separate electric (E) and magnetic (H) field MPE limits?

How may dangerous levels of carbon monoxide from an emergency generator be detected?

What does SAR measure?

Which insulating material commonly used as a thermal conductor for some types of electronic devices is extremely toxic if broken or crushed and the particles are accidentally inhaled?

What toxic material may be present in some electronic components such as high voltage capacitors and transformers?

Which of the following injuries can result from using high-power UHF or microwave transmitters?

~~~~End of "Just the Questions without the Answers"~~~~

Congratulations on Your Great Work!

With your Extra Class license you will be at the top of the ham world and ready to give your best to Amateur Radio! You have worked hard to earn this honor. Good luck and God speed!